Dr. Birdley Teaches Science:
Mysteries of the Earth

Featuring the Comic Strip

Middle and High School

Written and Illustrated by Nevin Katz

Incentive Publications, Inc.
Nashville, Tennessee

About the Author

Nevin Katz is an illustrator and curriculum developer who lives in Watertown, Massachusetts with his wife Melissa and son Jeremy.

Nevin majored in Biology at Swarthmore College and went on to earn his Master's in Education at the Harvard Graduate School of Education. He began developing curriculum as a student teacher in Roxbury, Massachusetts.

"Mr. Katz" taught biology, chemistry, and other sciences for eight years, in grades 6 through 11. He currently works as an online curriculum developer and facilitator at the Educational Development Center (EDC) in Newton, Massachusetts.

Nevin's journey with Dr. Birdley and the cast began in the summer of 2002, when he started authoring the cartoon and using it in his science classes. From there, he developed the cartoon strip, characters, and curriculum materials. After designing and implementing the materials, he decided to develop them further and organize them into this series of books.

Cover by Geoffrey Brittingham
Edited by Jill Norris
Science Editor Scott Norris

ISBN 978-086530-543-4

Copyright ©2009 by Incentive Publications, Inc., Nashville, TN. All Rights Reserved. The *Dr. Birdley* comic strip and all characters depicted in the comic strips, Copyright ©2009 by Nevin Katz. All rights reserved. The Dr. Birdley logo, Dr. Birdley™, Jaykes™, Dean Owelle™, Professor Lark™, Gina Sparrow™, and all prominent characters featured in this publication are trademarks of Nevin Katz.

No part of this publication may be reproduced, stored in a retrieval system, or transmitted in any form or by any means (electronic, mechanical, photocopying, or otherwise) without written permission from Incentive Publications, with the exception below:

Pages labeled with the statement Copyright ©2009 by Incentive Publications are intended for reproduction within the owner's classes. Permission is hereby granted to the purchaser for one copy of *Mysteries of the Earth* to reproduce these pages in sufficient quantities for meeting the purchaser's classroom needs only. Please include the copyright information at the bottom of each page on all copies.

1 2 3 4 5 6 7 8 9 10   12 11 10 09

PRINTED IN THE UNITED STATES OF AMERICA
www.incentivepublications.com
www.birdleymedia.com

TABLE OF CONTENTS

Contents

Teacher's Guide	5
Unit 1: The Fossil Record	11
Unit 2: Geologic Time	20
Unit 3: Origin of Life	33
Unit 4: Microbial Planet	42
Unit 5: Introducing Rocks	52
Unit 6: Sedimentary Rocks	63
Unit 7: Igneous Rocks and Volcanoes	77
Answer Key	89

Educational Objectives

Central Goal: To provide an overview of the major eras of life on Earth. To describe the processes that led to the formation of Earth's geological features

Chapter or Unit	Primary Objective(s)	Standards
1. The Fossil Record	Discuss how the principal of superposition and rock dating allow us to obtain information about Earth's history.	1, 9
2. Geologic Time	To provide an overview of geologic time. To explain how fossil evidence reflects changes in Earth's living things over time. To introduce the possible causes and the effects of mass extinctions.	1, 8, 10
3. Origin of Life	To introduce undersea vents as a plausible location for the origin of life. To show how microbes break down rocks and provide clues to life's origins.	2, 3, 10
4. Microbial Planet	To understand how the three domain scheme represents microbial diversity. To relate microbial biodiversity to the origins of the matter cycles.	2, 3, 10, 11, 13, 14, 15
5. Introducing Rocks	To compare and contrast igneous, sedimentary, and metamorphic rock. To understand the essential processes of the rock cycle.	4, 5, 11
6. Sedimentary Rocks	To illustrate how biological, chemical, and detrital sedimentary rocks are formed.	3, 4, 5, 6, 11
7. Igneous Rocks and Volcanoes	To distinguish between extrusive and intrusive igneous rocks. To explore the fundamental parts of a volcano. To examine evidence for seafloor spreading, which can be found near undersea volcanoes.	4, 5, 7, 11, 12

Dr. Birdley Teaches Science – Mysteries of the Earth

Relevant Frameworks

Grade 5-8: Earth, Space, and Life Science

The Fossil Record

1. Fossils provide important evidence of how life and environmental conditions have changed. (Ch. 1, 2)

Living Things

2. Most organisms are single cells; other organisms, including humans, are multicellular. (Ch. 3, 4)

3. Living organisms have played many roles in the Earth system, including affecting the composition of the atmosphere, producing some types of rocks, and contributing to the weathering of rocks. (Ch. 3, 4, 6)

Geological Processes

4. Land forms are the result of a combination of constructive and destructive forces. Constructive forces include crustal deformation, volcanic eruption, and deposition of sediment, while destructive forces include weathering and erosion. (Ch. 5, 6, 7)

5. Some changes in the solid Earth can be described as the "rock cycle." Old rocks at the Earth's surface weather, forming sediments that are buried, then compacted, heated, and often recrystallized into new rock. Eventually, those new rocks may be brought to the surface by the forces that drive plate motions, and the rock cycle continues (Ch. 5, 6, 7).

6. Water is a solvent. As it passes through the water cycle it dissolves minerals and gases and carries them to the oceans. (Ch. 6)

7. The Earth processes we see today, including erosion, movement of lithospheric plates, and changes in atmospheric composition, are similar to those that occurred in the past. (Ch. 7)

8. Earth history is also influenced by occasional catastrophes, such as the impact of an asteroid or comet. (Ch. 2)

Grade 9-12: Earth, Space, and Life Science

The Fossil Record

9. Geologic time can be estimated by observing rock sequences and using fossils to correlate the sequences at various locations. Current methods include using the known decay rates of radioactive isotopes present in rocks to measure the time since the rock was formed. (Ch. 1)

Origin of Life

10. Evidence for one-celled forms of life—the bacteria—extends back more than 3.5 billion years. The evolution of life caused dramatic changes in the composition of the Earth's atmosphere, which did not originally contain oxygen. (Ch. 2, 3, 4)

How Microbes Shaped the Earth

11. Interactions among the solid Earth, the oceans, the atmosphere, and organisms have resulted in the ongoing evolution of the Earth system. (Ch. 4, 5, 6, 7)

Geological Processes

12. We can observe some changes such as earthquakes and volcanic eruptions on a human time scale, but many processes such as mountain building and plate movements take place over hundreds of millions of years. (Ch. 7)

13. Movement of matter between reservoirs is driven by the Earth's internal and external sources of energy. These movements are often accompanied by a change in the physical and chemical properties of the matter. (Ch. 4)

14. Each element on Earth moves among reservoirs in the solid earth, oceans, atmosphere, and organisms as part of geochemical cycles. (Ch. 4)

Biodiversity

15. Biological classifications are based on how organisms are related. Organisms are classified into a hierarchy of groups and subgroups based on similarities which reflect their evolutionary relationships. (Ch. 4)

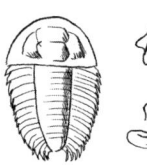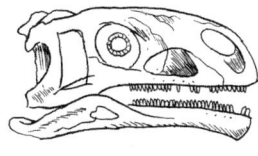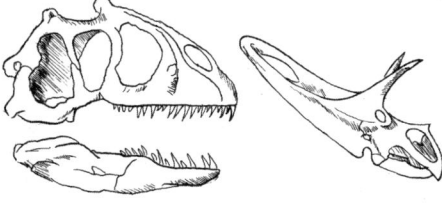

Overview of Source Cartoons

Cartoon	Central Concept	Challenge Level	Related Topics
The Fossil Record	The principle of superposition and radiometric dating are useful methods for learning about geologic time.	2	Sedimentary Rock
Radiometric Dating	Radiometric dating uses the half life and decay rates of isotopes within igneous rocks to find the age of fossils.	3	Isotopes
Snapshots in Time	Fossils reflect changes in Earth's biodiversity over time.	2	Fossil Formation
The Trilobite Case	Trilobites disappeared during the Permian Extinction, but they have left well-preserved fossils.	1	Natural Disasters
Geologic Time	Whereas microbes have been around for the majority of Earth's history, humans have appeared only recently.	1	Precambrian Time
Origin of Life	The origin of life may have occurred in undersea vents.	2	Geothermal Energy
Living Fossils	The snottites contain microbes that carve through cave walls with sulfuric acid and resemble the first living things.	2	Chemical Weathering of Rocks
Three Domains	Microbes are more diverse than all other forms of life because of their genetics, metabolism, and habitats.	2	Biogeochemical Cycles
Biogeochemical Cycles	The biodiversity of microbes allow them to metabolize a vast range of chemicals, resulting in biogeochemical cycles.	2	Chemical Bonding
Types of Rocks	The three major rock types—igneous, metamorphic, and sedimentary—each have different origins and characteristics.	2	Plate Tectonics
Transforming Rocks	Any rock could become any of the three major rock types.	2	Rock Cycle
Owelle's Rock	Detrital sedimentary rocks can form from weathering, compaction, and cementation.	1	Erosion
White Sands	The white sands are made of chemical sedimentary rock.	3	Evaporation
A Lesson from a Snail	Chalk is made of calcite, which is composed of the fossils of microbes like foraminifera and other marine organisms.	2	Protists
Rocks Above and Below	Extrusive igneous rock forms above ground, while sedimentary rock forms below ground.	2	Volcanoes
Anatomy of a Volcano	Composite volcanoes are built up by magma rising up through Earth's crust, forming layers of rock and ash.	2	Plate Tectonics
Deep Sea Smoker	Evidence for seafloor spreading can be found at undersea volcanoes known as "black smokers."	3	Chemosynthetic Bacteria

Dr. Birdley Teaches Science – Mysteries of the Earth

Teacher's Guide

Contents

Introduction 6

The Source Cartoon 7

The Cartoon Profile 8

Assignments and Assessments 9

Sample Lesson Plan 10

 THE SOURCE CARTOON

The Source Cartoon

The *Source Cartoon* explains the central concepts of the unit. It is usually one to two pages in length. Expect to find the following in a source cartoon:

- A central idea with supporting details
- Visual images related to the topic being presented
- Explanations of science concepts
- Several characters who explain the information to each other or to the reader

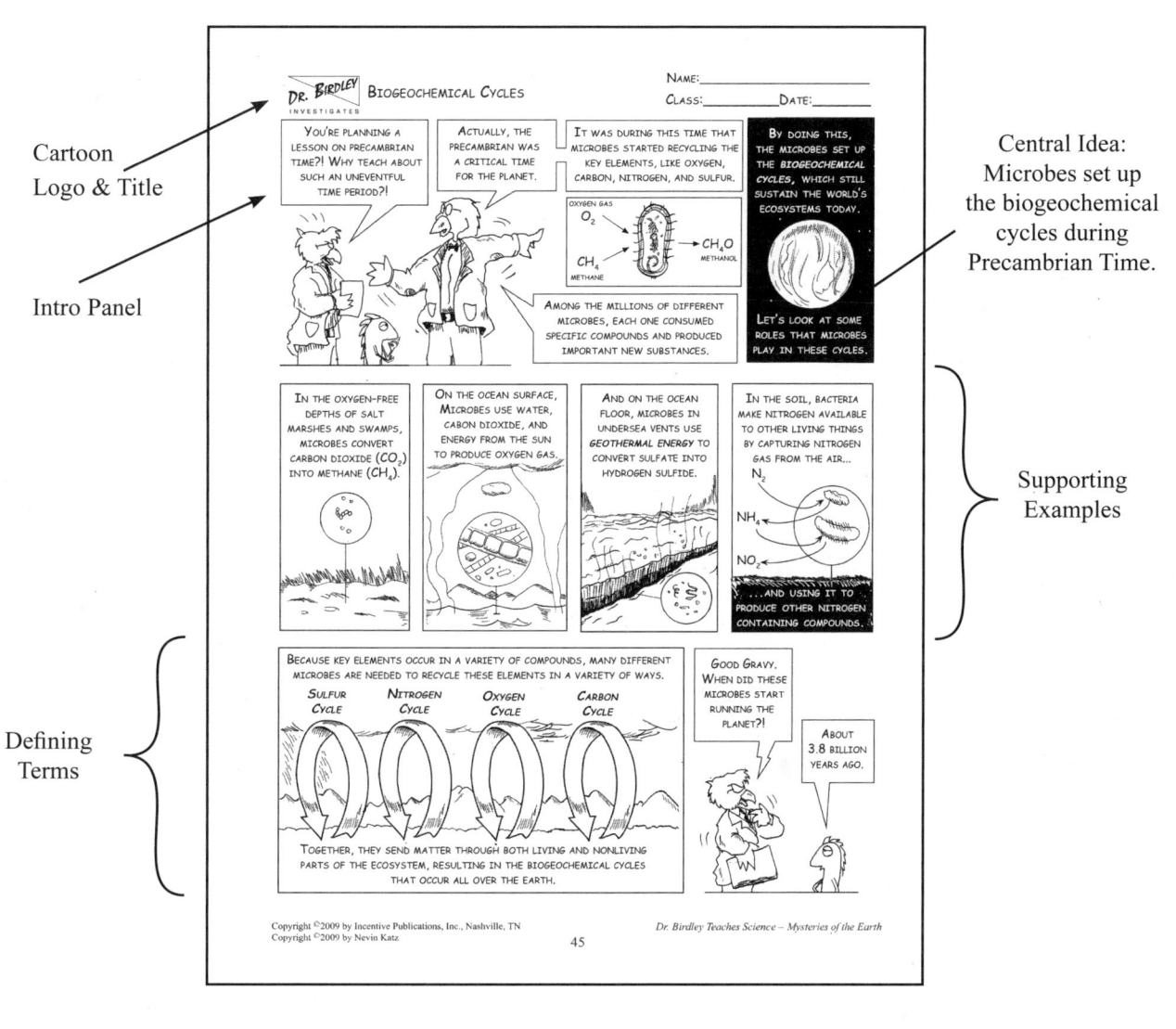

Dr. Birdley Teaches Science – Mysteries of the Earth

The Cartoon Profile

The *Cartoon Profile*, which outlines a source cartoon's science content, is useful for planning or teaching a lesson. Central elements include:

- The objectives in the cartoon, which tie back to the national standards.

- The "questions for discussion" below the image, which are useful for getting students engaged and checking for understanding.

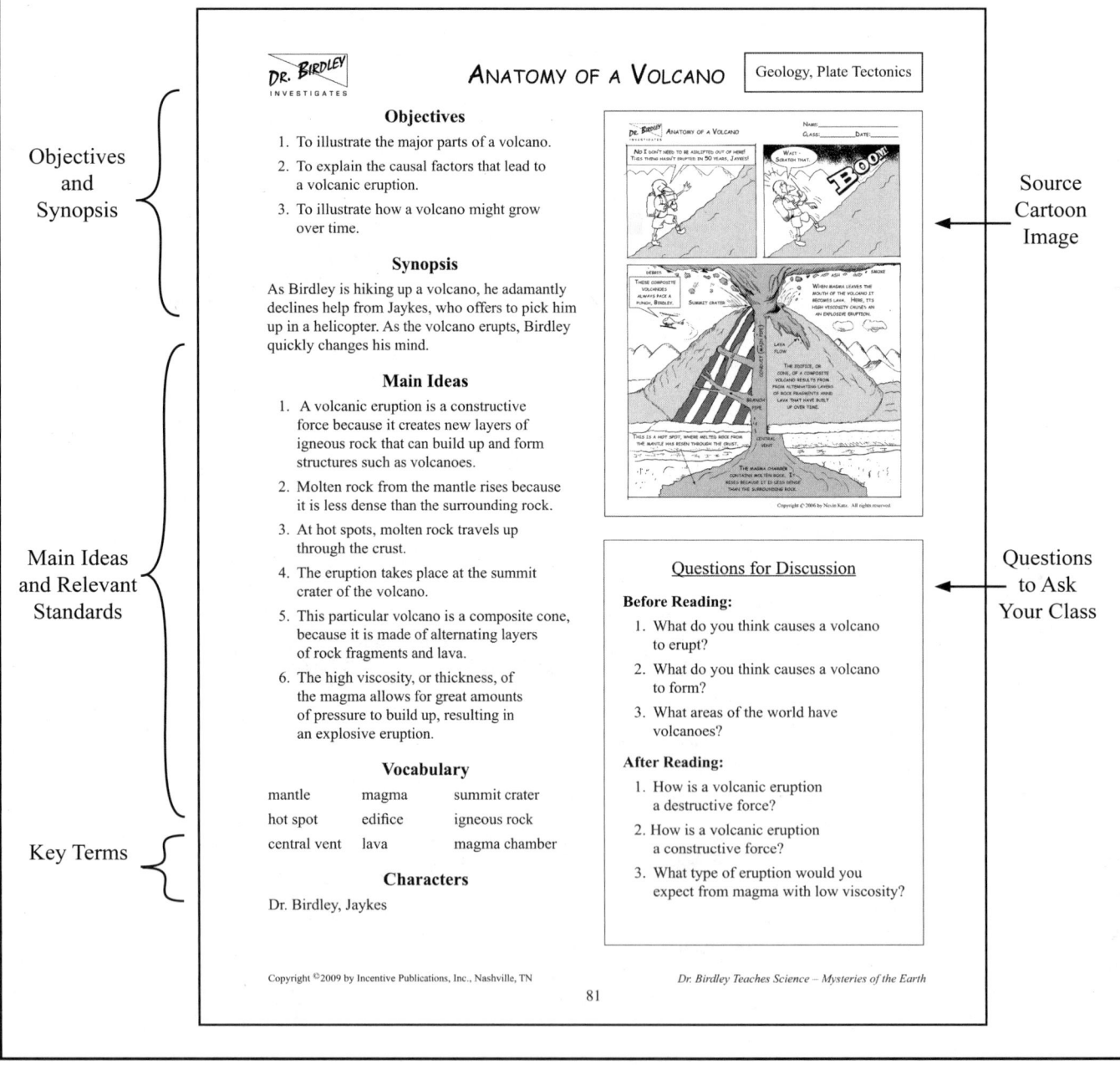

Assignments and Assessments

Supplementary assignments help students build comprehension of the central ideas in the cartoons. The assignments are based on the central points of the source cartoon(s) and cartoon profile(s). Five of the major resources and a quiz are pictured below:

Study Questions

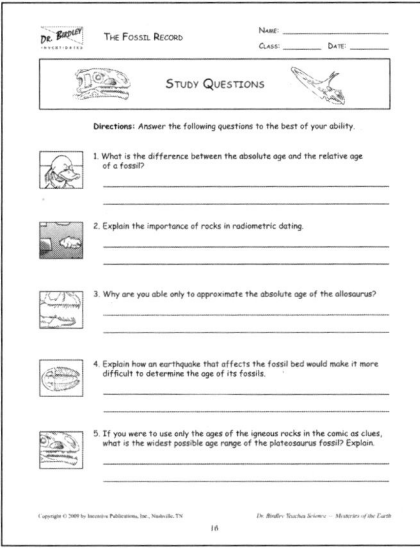

Visual Exercises

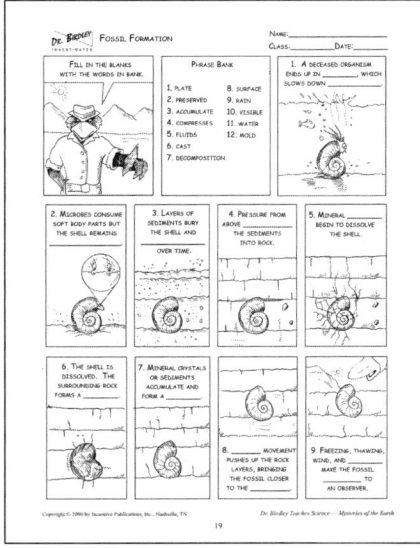

Graphic Organizer

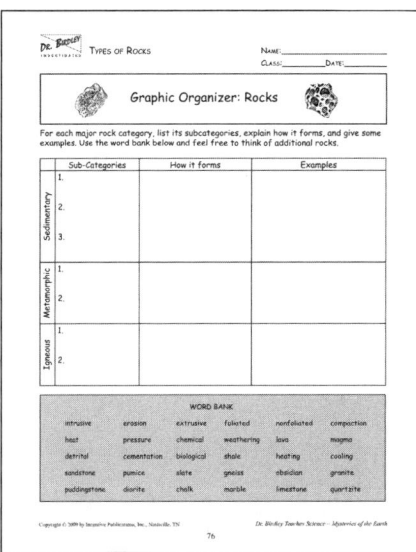

Vocabulary Build-up

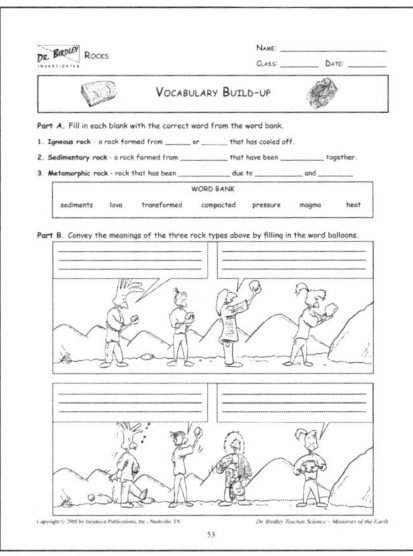

Background Article

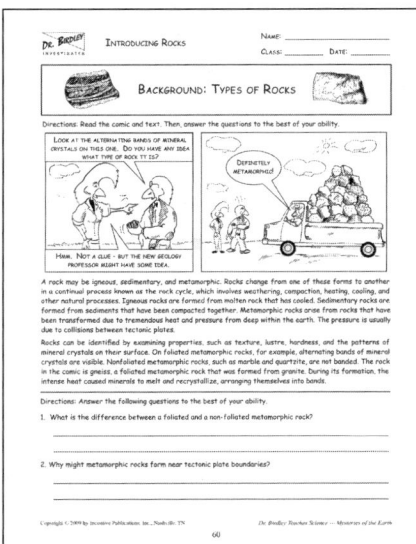

Quizzes

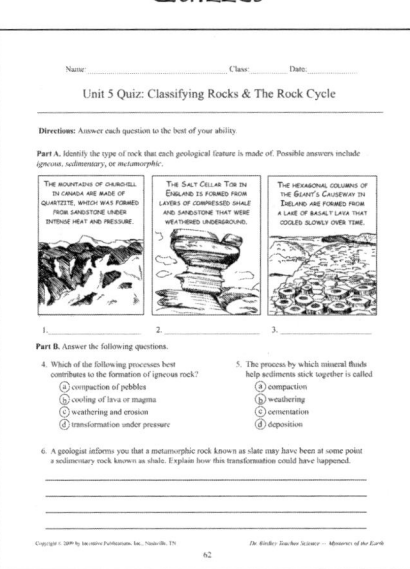

Sample Lesson Plan

This lesson plan uses the source cartoon "The Fossil Record" (page 12) and the related study questions (p. 16) to introduce methods for analyzing fossils. A visual exercise (page 19) is used to illustrate how some fossils form. This format is transferable to a range of different lessons.

Warming up the brain!

Lesson Objective: To understand how fossils are used to learn about the history of life on Earth and examine the process of fossil formation.

A. *Warm-up:* Students list three to five examples of fossils that they have seen or heard of. Students then list three body parts that fossilize easily and explain why.

B. *Sharing ideas:* Students share their answers and take turns going to the board to draw examples of fossils.

C. *Vocabulary:* Students learn the terms *superposition* and *radiometric dating* and use them in sentences.

Reading and Discussion

A. *Introducing the Cartoon:* The teacher leads a discussion on the *Before Reading* questions from the cartoon profile.

B. *Classwide Reading:* Several student volunteers read the cartoon aloud. Others read while underlining key words.

C. *Discussion:* The teacher leads a discussion on the *After Reading* questions from the cartoon profile.

D. *Reading in Pairs:* Students read again in pairs, searching for the main ideas or particular key terms in the comic.

Practice and Application

A. *Independent Practice:* Students are instructed to complete the study questions.

B. *Pairs:* Students review their answers to the activity in pairs.

C. *Report Out:* Pairs justify and explain one or more answers.

D. *Teacher Feedback:* The teacher then offers his/her insight on how accurate the answers are.

E. *Timeline:* Students complete the visual exercise on fossil formation and repeat the above steps B–D (page 19).

Dr. Birdley Teaches Science – Mysteries of the Earth

Unit 1: The Fossil Record

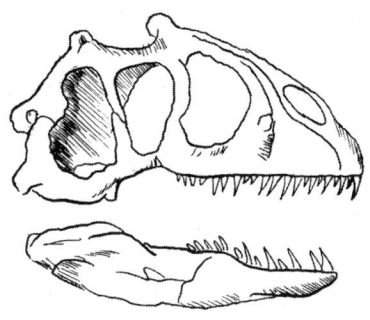

Contents

Source Cartoon: The Fossil Record 12

Source Cartoon: Radiometric Dating 13

Cartoon Profiles 14

Study Questions 16

Background 18

Visual Exercise 19

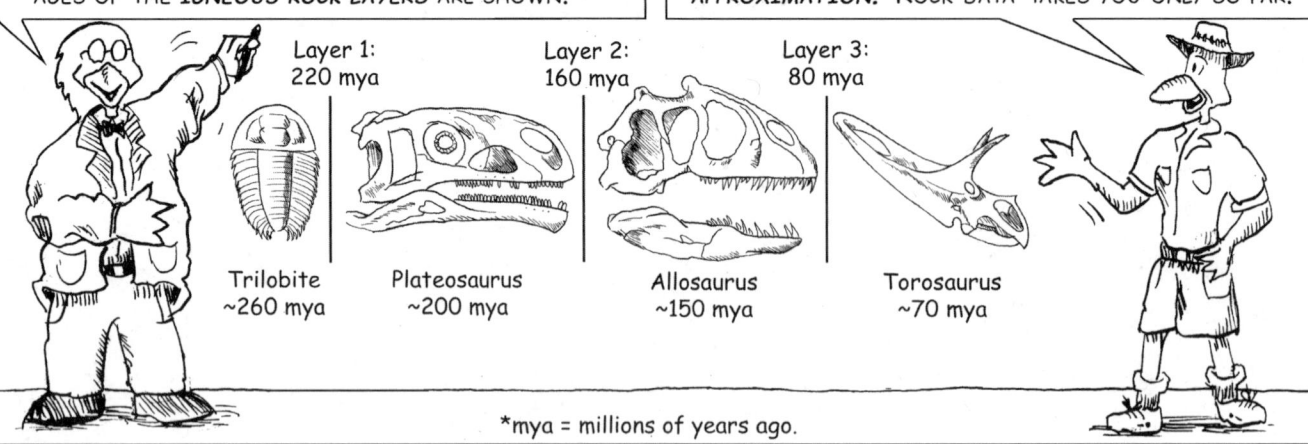

RADIOMETRIC DATING

NAME:_____

CLASS:_____ DATE:_____

Copyright ©2009 by Incentive Publications, Inc., Nashville, TN
Copyright ©2009 by Nevin Katz

Dr. Birdley Teaches Science – Mysteries of the Earth

THE FOSSIL RECORD

Fossils and Geologic Time

Objectives

1. To explain how scientists learn about the relative age of fossils using the principle of superposition.
2. To explain that scientists can learn about the relative age of fossils using radiometric dating.

Synopsis

Dr. Birdley stands atop a fossil bed explaining the principle of superposition. The Earth below him contains skulls of three dinosaurs and a fossilized trilobite. Birdley then explains radiometric dating with Eric Seagull and Arthur Grouse. A timeline is used to illustrate the order in which the dinosaurs and trilobite appeared in Earth's history. Meanwhile, Eric discusses the limitations of the fossil record.

Main Ideas

1. The principle of superposition can be used to learn about the relative ages of fossils.
2. Radiometric dating can be used to find the absolute (numerical) age of igneous rocks.
3. Because layers with fossils are only made of sedimentary rock, your best strategy is to find the ages of igneous rock layers above and below the fossil-containing layer.
4. The fossil's age falls within a range between the ages of the two igneous rock layers.
5. You can use this date range to approximate the age of the fossil.

Vocabulary

superposition   radiometric dating
relative age   absolute age
igneous rock   sedimentary rock

Characters

Dr. Birdley, Eric Seagull, Arthur Grouse

Questions for Discussion

Before Reading:

1. What can fossils tell us about the past?
2. What are some examples of living things that are extinct?
3. What are the various types of fossils that exist?

After Reading:

1. How could a geological event such as an earthquake interfere with a scientist's ability to use the principle of superposition?
2. Why is Eric able to use only approximations?

Radiometric Dating

Fossils & Geologic Time

Objectives

1. To explain how the age range of a fossil can be determined using radiometric dating.
2. To illustrate an example of rock layers in which a fossil has been found.

Synopsis

Having brought the femur bone of an apatasaurus back to his lab, Eric Seagull explains to Dr. Birdley how he determined the age of the fossil using radiometric dating.

Main Ideas

1. Radiometric dating is used to determine or approximate the absolute age of a fossil.
2. Uranium is useful for finding the age of dinosaur fossils because it has a long half life.
3. Radiometric dating works best with igneous rocks.
4. While fossils are typically found in sedimentary rocks, you can infer their ages by determining the ages of igneous rock layers above and below the fossil.
5. The amount of uranium that has decayed can be used to calculate the age of the sample.

Vocabulary

radiometric dating	half life	uranium
carbon-14	igneous	sedimentary
fossil	femur	apatosaurus

Characters

Dr. Birdley, Eric Seagull, subway passengers

Teacher's Note

Carbon-14 is another isotope that is used to date rocks. However, because of its short half-life of 5,730 years, it can only be used to determine the age of human artifacts and recent fossils.

Questions for Discussion

Before Reading:

1. Pick an object that you own at home. How can you determine how old the object is?
2. Why is it useful to know the age of a fossil?
3. What fossils are you most interested in?

After Reading:

1. What steps are involved in radiometric dating?
2. Suppose there were no igneous rock layers above or below the fossil layer. How would this have affected Eric's ability to find the age of the bone?

THE FOSSIL RECORD

NAME: _____

CLASS: _____ DATE: _____

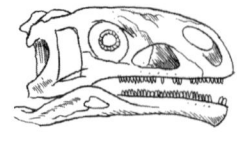 ## STUDY QUESTIONS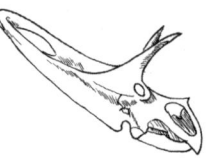

Directions: Answer the following questions to the best of your ability.

1. What is the difference between the absolute age and the relative age of a fossil?
_____
_____

2. Explain the importance of rocks in radiometric dating.
_____
_____

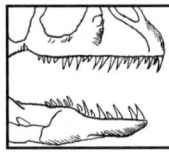
3. Why are you able to only approximate the absolute age of the allosaurus?
_____
_____

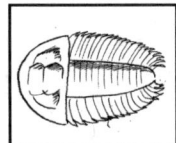
4. Explain how an earthquake that affects the fossil bed would make it more difficult to determine the age of its fossils.
_____
_____

5. If you were to use only the ages of the igneous rocks in the comic as clues, what is the widest possible age range of the plateosaurus fossil? Explain.
_____
_____

Copyright ©2009 by Incentive Publications, Inc., Nashville, TN

Dr. Birdley Teaches Science – Mysteries of the Earth

 **RADIOMETRIC DATING**

NAME: _____
CLASS: _____ DATE: _____

STUDY QUESTIONS

Directions: Answer the following questions to the best of your ability.

1. Explain the role of uranium-235 in radiometric dating.
   _____
   _____

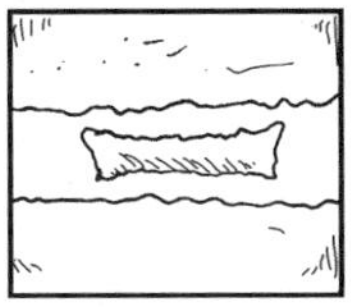

2. Why is it fortunate that the rock layer with the fossil was in between two sedimentary rock layers?
   _____
   _____

3. A rock that originally had 70 g of uranium-235 now has only 35 g of uranium-235. How long has the rock been around? Explain your answer.
   _____
   _____

4. Why did Eric describe the age of the fossil using a time frame as opposed to a single date?
   _____
   _____

5. Carbon-14, which has a half-life of 5,730 years, is also used in radioactive dating. Would this be more or less effective with a dinosaur bone? Why?
   _____
   _____

Copyright ©2009 by Incentive Publications, Inc., Nashville, TN

Dr. Birdley Teaches Science – Mysteries of the Earth

THE FOSSIL RECORD

NAME: _____

CLASS: _____ DATE: _____

Background: Searching for Fossils

THANKS TO **EROSION**, SOME FOSSILS IN THIS SEDIMENTARY ROCK ARE NOW VISIBLE!

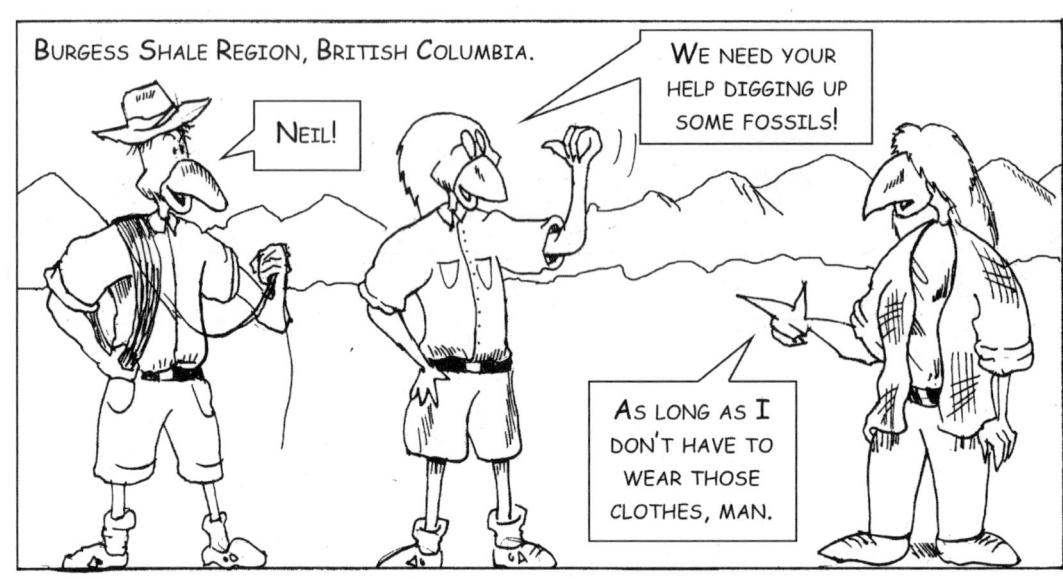

Fossils are great storytellers of geologic time. A fossil provides clues to what a species of the past looked like, when it lived, and how it behaved. Some organisms leave body fossils, such as whole dinosaur skeletons. Others leave only trace fossils: remnants of the past such as egg shells, footprints, or preserved leaves.

After an organism dies, a fossil may form if its hard body parts end up buried in sedimentary rock. Over many years, the layers above the fossil are sometimes worn away by erosion, causing the fossil to become visible to an observer. You can determine the relative ages of fossils by using the principle of superposition, which states that the lower fossil layers are typically older than those above them. This logic may not be enough for fossil beds that have undergone earthquakes, which can overturn fossil layers.

To find out more about a fossil's age, you need to use radioactive dating to find the ages of igneous rocks that were above and below it. Igneous rocks are known to contain radioactive elements, such as Uranium-235, which decay into other elements over time. The half-life of uranium-235, which is 174 million years, is the amount of time it takes for half of a given sample to decay into lead. Knowing the ratio of uranium to lead in a given rock can give scientists insight into how old the rock is, and obtain clues to the fossil's age.

Directions: Answer the following questions to the best of your ability.

1. Explain how a fossil buried in sedimentary rock can become visible.

_____

_____

2. Explain why radiometric dating is important in examining fossils.

_____

_____

DR. BIRDLEY INVESTIGATES — FOSSIL FORMATION

NAME:_____
CLASS:_____ DATE:_____

FILL IN THE BLANKS WITH THE WORDS IN BANK.

PHRASE BANK

1. PLATE
2. PRESERVED
3. ACCUMULATE
4. COMPRESSES
5. FLUIDS
6. CAST
7. DECOMPOSITION
8. SURFACE
9. RAIN
10. VISIBLE
11. WATER
12. MOLD

1. A DECEASED ORGANISM ENDS UP IN _____, WHICH SLOWS DOWN _____.

2. MICROBES CONSUME SOFT BODY PARTS BUT THE SHELL REMAINS _____.

3. LAYERS OF SEDIMENTS BURY THE SHELL AND _____ OVER TIME.

4. PRESSURE FROM ABOVE _____ THE SEDIMENTS INTO ROCK.

5. MINERAL _____ BEGIN TO DISSOLVE THE SHELL.

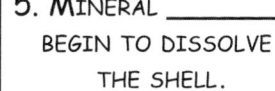
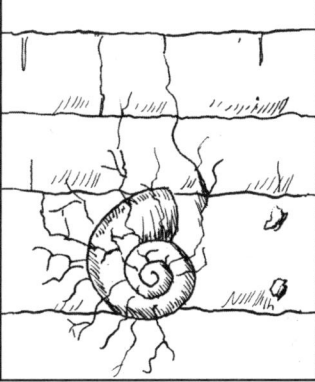

6. THE SHELL IS DISSOLVED. THE SURROUNDING ROCK FORMS A _____.

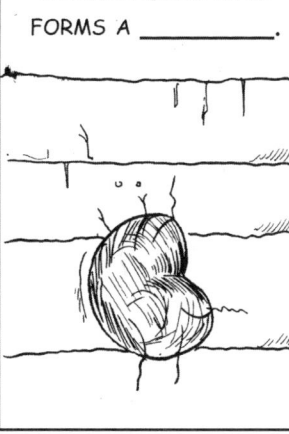

7. MINERAL CRYSTALS OR SEDIMENTS ACCUMULATE AND FORM A _____.

8. _____ MOVEMENT PUSHES UP THE ROCK LAYERS, BRINGING THE FOSSIL CLOSER TO THE _____.

9. FREEZING, THAWING, WIND, AND _____ MAKE THE FOSSIL _____ TO AN OBSERVER.

Unit 2: Geologic Time

Contents

Source Cartoon: Geologic Time 21

Cartoon Profile 22

Study Questions 23

Source Cartoon: Snapshots in Time 24

Source Cartoon: The Trilobite Case 25

Cartoon Profiles 26

Study Questions 28

Background 30

Visual Exercise 31

Quiz 32

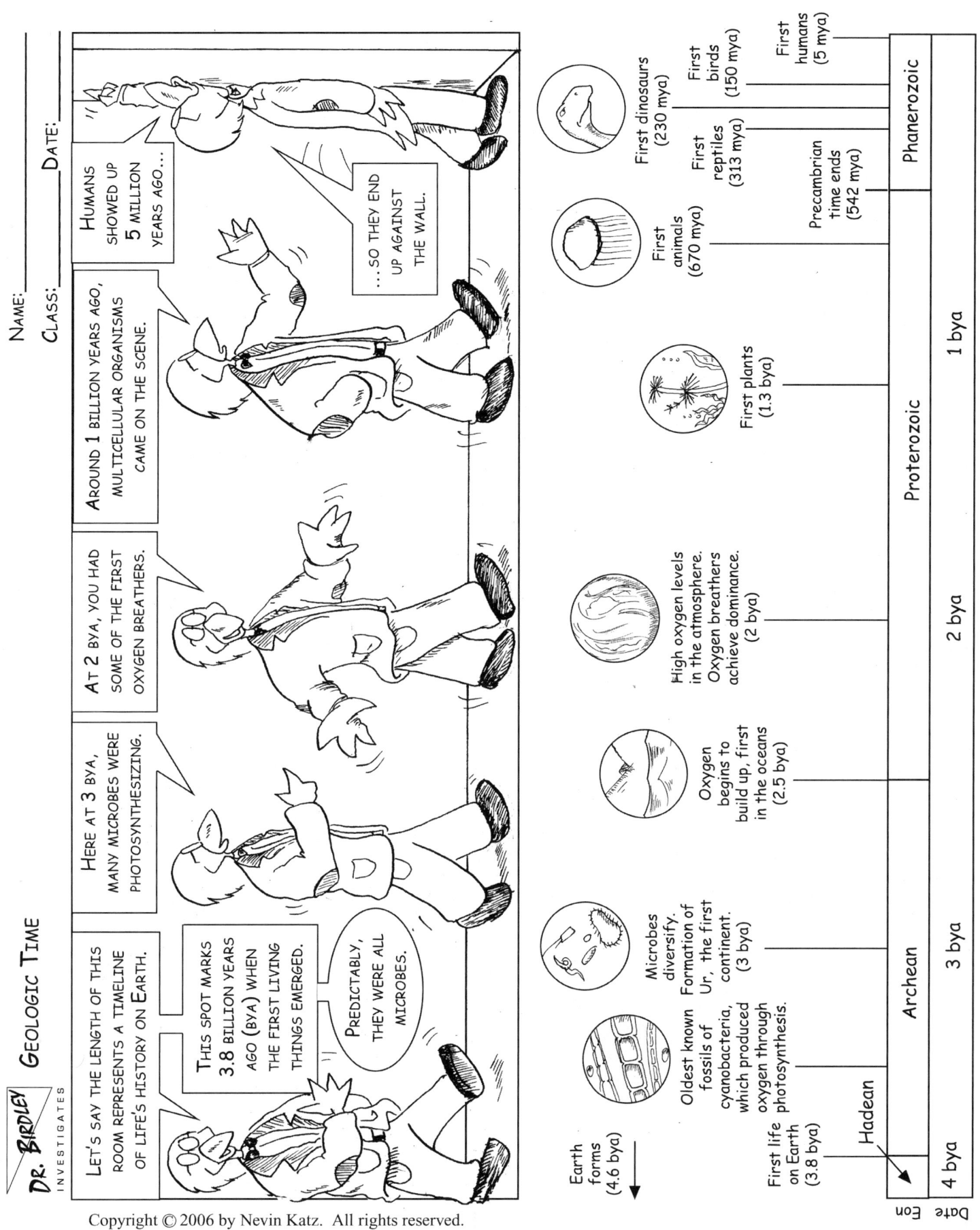

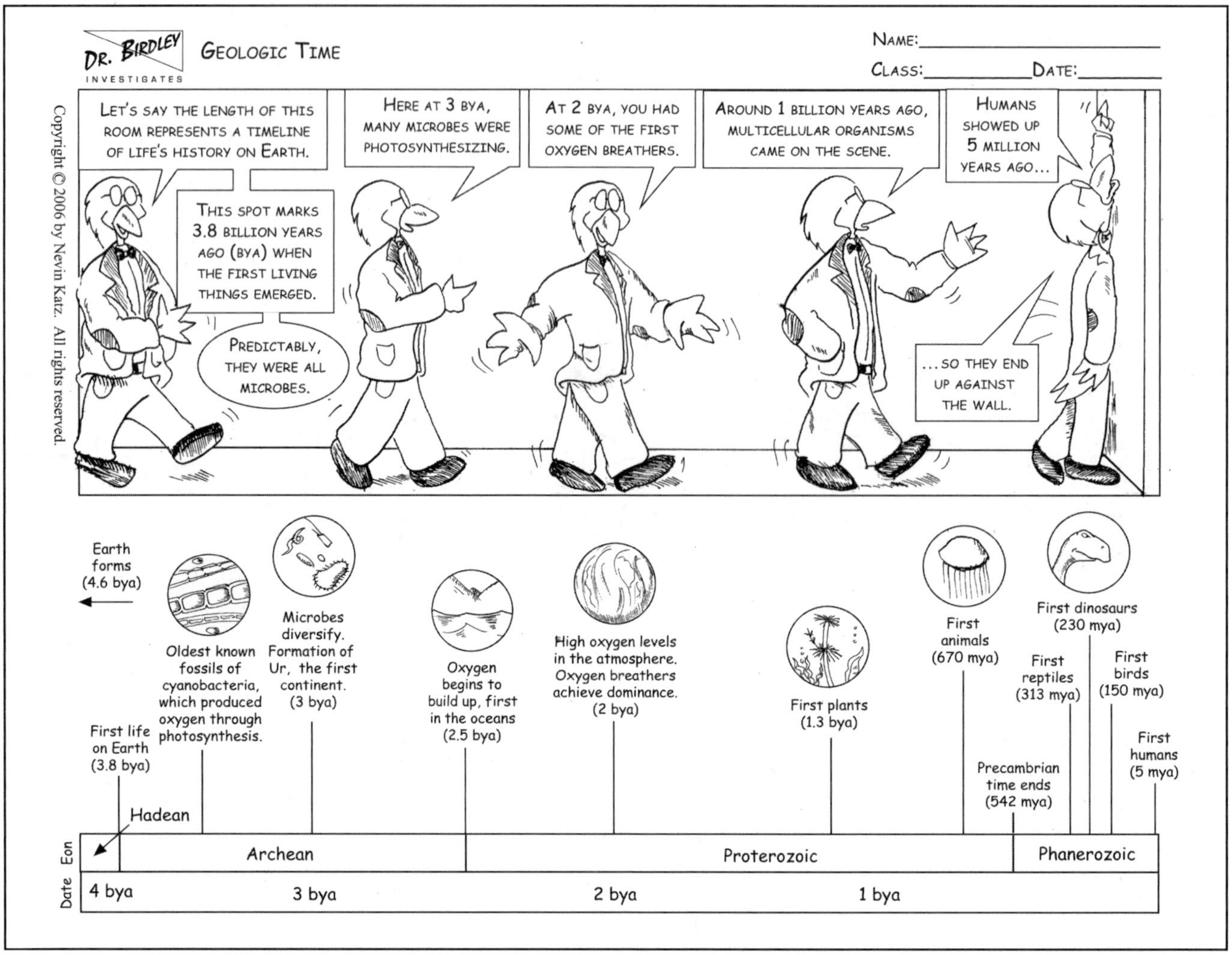

Geologic Time

Geologic Time

Objectives

1. To illustrate the order of key natural events in the history of life on Earth.
2. To illustrate the point at which humans emerged on Earth over the course of geologic time.

Synopsis

Dr. Birdley uses the length of a room to represent geologic time, and discusses when several key events occurred in Earth's history.

Main Ideas

1. Microbial life dominated Earth for 3.23 billion years (3.8 bya – 0.67 bya).
2. While Earth formed 4.6 bya, life is thought to have originated 3.8 bya.
3. The first humans emerged 5 million years ago, which is very recent on the timeline.
4. Precambrian time stretches from 4.6 bya to 0.542 bya.

Vocabulary

hadean   archean   proterozoic
Precambrian   microbes   cyanobacteria

Question before Reading:

List ten key events in Earth's history and put them in chronological order.

Questions after Reading:

1. What surprised you about this timeline?
2. What additional events would you add?

GEOLOGIC TIME

NAME: _____

CLASS: _____ DATE: _____

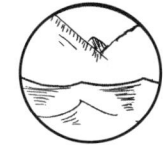 

STUDY QUESTIONS

Directions: Answer the following questions to the best of your ability.

1. What does the comic tell you about when humans emerged over the course of Earth's history?

   _____

   _____

2. According to the timeline, what type(s) of organism(s) have been on Earth for the longest period of time? What is the evidence for this?

   _____

   _____

3. Why were cyanobacteria and other photosynthetic organisms important in the history of the Earth?

   _____

   _____

4. What change in Earth's atmosphere occurred before animals emerged? Why was this change important?

   _____

   _____

5. According to the timeline, do dinosaurs and plants show up in the early or later part of Earth's history? Explain.

   _____

   _____

SNAPSHOTS IN TIME

NAME:_____
CLASS:_____ DATE:_____

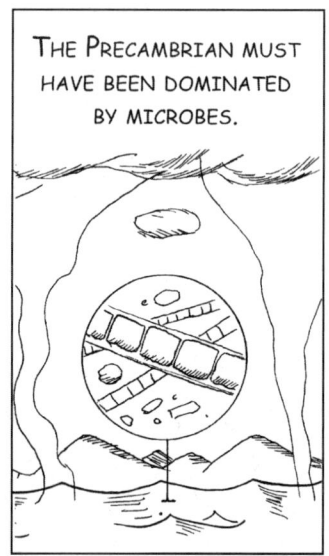

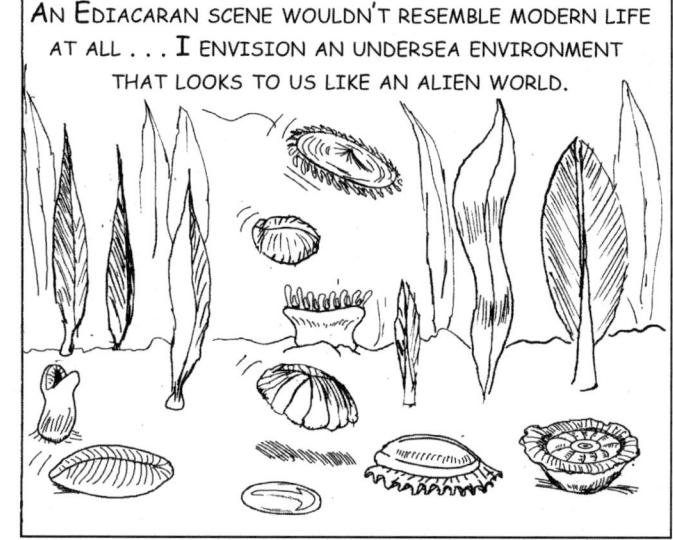

Snapshots in Time

Geologic Time

Objectives

1. To illustrate how fossils reflect changes in the types of living things that populate the Earth.
2. To illustrate how fossils reflect shifting levels of biodiversity over time.

Synopsis

As Gina travels to various countries to dig up fossils, she visualizes what a typical scene from that time period may have looked like.

Main Ideas

1. The scarce fossil evidence and presence of microfossils in early fossil layers suggests that Precambrian life was predominantly microbial.
2. Fossils from the Ediacaran Era reflect the first major emergence of complex multicellular life.
3. Fossils indicate that the Cambrian Explosion was a period of rapid diversification, in which organisms from all the major animal phyla emerged.
4. Fossils can give us insight as to what living things existed in particular time periods.

Vocabulary

Cambrian   Ediacaran   Precambrian Time
microfossils   microbes   phyla

Characters

Gina

Teacher's Note

The Cambrian Period was the first part of the Paleozoic Era. Although an explosion in biodiversity ensued during this time frame, much of it was lost by the end of the Paleozoic Era due to the Permian Extinction.

Questions for Discussion

Before Reading:

1. When do you think the first animals appeared?
2. What were the first living things?
3. How can fossils help us investigate this question?

After Reading:

1. Why do you think life during the Precambrian was mainly microbial?
2. What may have caused the sudden increase in biodiversity during the Cambrian Period.
3. Ask students to describe what they think an ecosystem from the Cambrian Period may have looked like.

Copyright ©2009 by Incentive Publications, Inc., Nashville, TN

Dr. Birdley Teaches Science – Mysteries of the Earth

THE TRILOBITE CASE

Geologic Time

Objectives

1. To explain what may have caused the Permian Extinction.
2. To introduce the idea that the Permian Extinction had wide-ranging effects on the ecosystems of the planet.

Synopsis

Detective Benedict, Eric Seagull, and Dr. Birdley find a trilobite fossil. As Detective Benedict speculates about the cause of the trilobite's demise, Eric Seagull points out that the Permian Extinction ended the time of the trilobites and 95% of other species. After pondering the cause of the Permian Extinction, Dr. Birdley goes back in time to the end of the Pemian Period in order to learn more.

Main Ideas

1. Trilobites lived during the Paleozoic Era.
2. Trilobites are ancestors of the horseshoe crab.
3. Trilobites became extinct due to the Permian Extinction, which wiped out 95% of all Earth's species.
4. The cause of the Permian Extinction may have been volcanism, climate change, continental drift, meteorites, and glaciers, but nobody knows for sure.

Vocabulary

Paleozoic Era   trilobite   Permian Extinction
volcanism   meteorites   continental drift
exoskeleton   climate

Characters

Dr. Birdley, Eric Seagull, Eggs Benedict, Trilobite

Questions for Discussion

Before Reading:

1. What do you know about extinctions?
2. What are examples of living organisms from the past that are now extinct?
3. What types of animals form the best fossils? Why?
4. Where have you seen horseshoe crabs or trilobite fossils?

After Reading:

1. If you could go back in time to learn about the life of the past, what period would you visit? Why?
2. What clues in the fossil record may indicate an extinction? Try drawing what fossil layers may look like.

Copyright ©2009 by Incentive Publications, Inc., Nashville, TN

Dr. Birdley Teaches Science – Mysteries of the Earth

 SNAPSHOTS IN TIME

NAME: _____

CLASS: _____ DATE: _____

 STUDY QUESTIONS

Directions: Answer the following questions to the best of your ability.

1. Why does Gina think that the majority of the Precambrian Period was dominated by micro-organisms?

   _____

   _____

2. Why was the Ediacaran Period significant?

   _____

   _____

3. Based on fossil evidence, how did life on Earth change between the Ediacaran Period and the Cambrian Period?

   _____

   _____

4. Explain why the fossils from the Cambrian Period reflected a "Cambrian Explosion."

   _____

   _____

5. If the last panel is any indication, where on Earth did life primarily exist during the Precambrian? Why do you think this was the case?

   _____

   _____

THE TRILOBITE CASE

NAME: _____

CLASS: _____ DATE: _____

 STUDY QUESTIONS

Directions: Answer the following questions to the best of your ability.

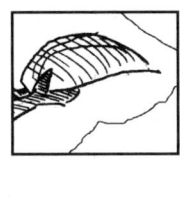
1. What part of the trilobite is preserved? Why?
   _____
   _____

2. What is the result of an extinction event?
   _____
   _____

3. Why did the Permian Extinction have such a dramatic impact on the biosphere?
   _____
   _____

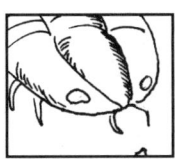
4. How might climate change cause an extinction?
   _____
   _____

5. What do you think is the most likely cause of the Permian Extinction? Why?
   _____
   _____

Copyright ©2009 by Incentive Publications, Inc., Nashville, TN

Dr. Birdley Teaches Science – Mysteries of the Earth

GEOLOGIC TIME

NAME: _____

CLASS: _____ DATE: _____

Background: Fossils Reveal Mysteries of Geologic Time

While fossils from Precambrian Time are rare, the world's fossils tell us a lot about the Paleozoic, Mesozoic, and Cenozoic eras that followed. Animal fossils tell us that the Paleozoic was a time when animal diversity exploded in the seas. It was also a time when insects, fish, reptiles, land plants, and vertebrates emerged for the first time. Fossils from the Mesozoic Era show that dinosaurs roamed the Earth at this time. Plants also evolved in response to insect behavior, sporting attractive flowers that encouraged pollination. And as small mammals emerged, so did the first birds. In the Cenozoic Era, mammal life diversified further and primates evolved into the first humans. Events such as these were revealed to us by fossils.

Directions: Answer the following questions to the best of your ability.

1. Give an example of how a fossil may help us learn about the past.

   _____

   _____

2. Explain how a plant with flowers may have a reprductive advantage over a plant without flowers..

   _____

   _____

3. Why do you think fossils from time periods after the Precambrian are more common?

   _____

   _____

History of Living Things

Name:_____
Class:_____ Date:_____

Describe at least three major events in each of the time periods below. Use the pictures and the phrase bank on the right for ideas.

Phrase Bank

- DINOSAURS
- MAINLY MICROBES
- FIRST FLOWERING PLANTS
- TRILOBITES
- FIRST VERTEBRATES
- FIRST LAND PLANTS
- FIRST ANIMAL TRACES
- FIRST SKELETAL ELEMENTS
- MAMMALS DIVERSIFY
- PANGAEA BREAKS APART
- LARGE MAMMALS
- FIRST FISH
- HUMAN EVOLUTION
- FIRST REPTILES
- FIRST MAMMALS
- FIRST LIVING THINGS
- FIRST INSECTS

1. Precambrian (4.6 bya - 542 mya)

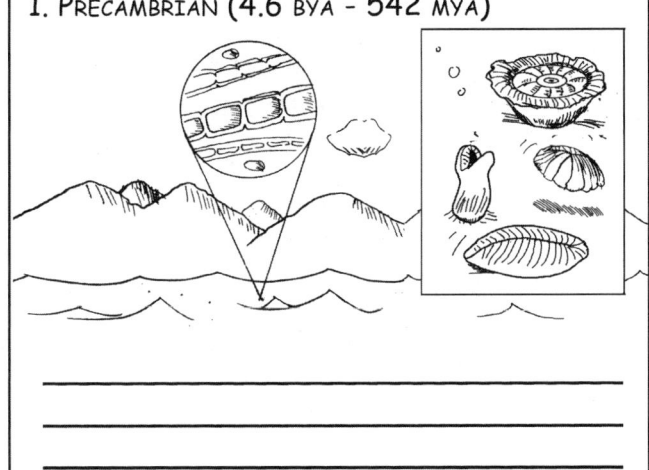

_____
_____
_____

2. Paleozoic (542 mya - 251 mya)

_____
_____
_____

3. Mesozoic (251 - 65.5 mya)

_____
_____
_____

4. Cenozoic (65.5 mya - Present)

_____
_____
_____

Copyright ©2009 by Incentive Publications, Inc., Nashville, TN

Dr. Birdley Teaches Science – Mysteries of the Earth

Name:_____ Class:_____ Date:_____

Unit 1 and 2 Quiz: Fossils and Geologic Time

Directions: This quiz tests your knowledge of the chapters' comics, background article, and visual exercises. Answer the following questions to the best of your ability.

1. Which of the following groups of organisms has existed for the longest amount of time on Earth?
   a) animals
   b) microbes
   c) plants
   d) humans

2. The best way to find the absolute age of a fossil is to
   a) use the principle of superposition
   b) examine the surrounding landscape
   c) determine the type of organism that left it behind
   d) use radiometric dating on neighboring rocks

3. Which of the following organisms lived during the Paleozoic Era?
   a) flowers
   b) dinosaurs
   c) trilobites
   d) large mammals

4. Which of the following is an important event that occurred during Precambrian Time?
   a) break-up of Pangaea
   b) oxygenation of Earth's atmosphere
   c) formation of the himalayas
   d) the emergence of the first dinosaurs

5. During fossil formation, a cast is
   a) the actual fossil that is dissolving
   b) a replica of the fossil that forms from mineral crystals and sediments
   c) a cavity in the shape of the fossil that is left after the fossil deteriorates
   d) the pressure from above that causes sediments to harden into rock

6. The Cambrian period of the Paleozoic is best known for a/an
   a) emergence of the first primates
   b) diversification of land plants
   c) explosion of animal biodiversity
   d) mass extinction caused by climate change

7. Name one global or regional catastrophe and discuss how this event could cause a mass extinction.

_____

_____

_____

_____

Unit 3: Life's Origins

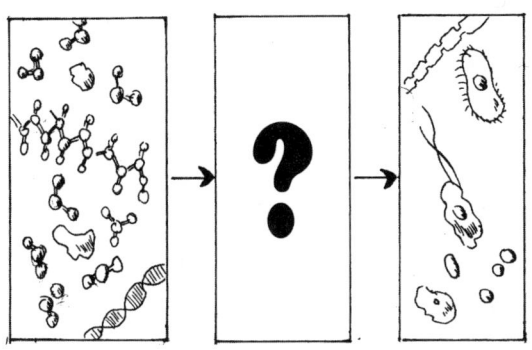

Contents

Source Cartoon: Origin of Life 34

Source Cartoon: Living Fossils 35

Cartoon Profiles 36

Study Questions 38

Background 40

Visual Exercise 41

DR. BIRDLEY INVESTIGATES: Origin of Life

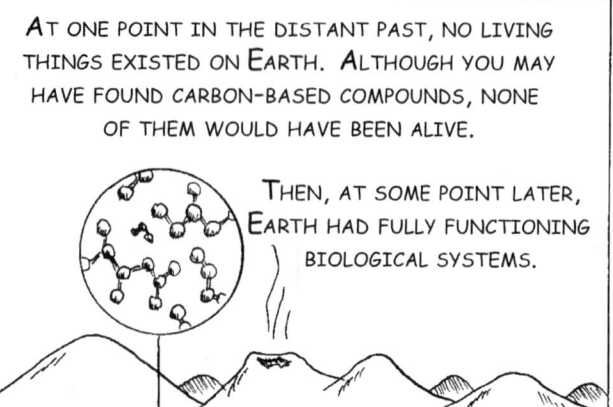

At one point in the distant past, no living things existed on Earth. Although you may have found carbon-based compounds, none of them would have been alive.

Then, at some point later, Earth had fully functioning biological systems.

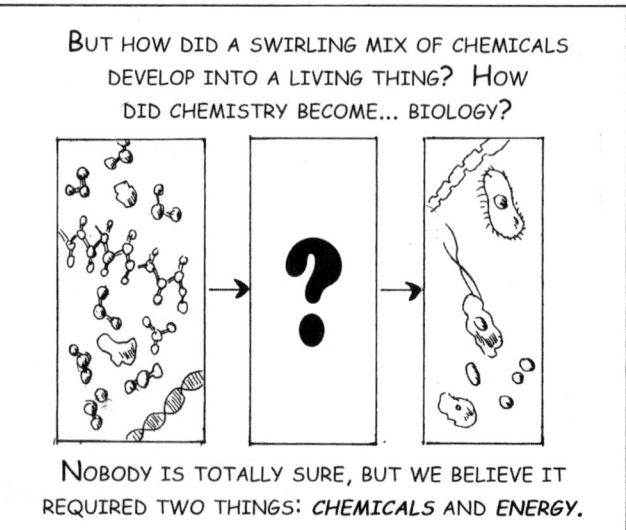

But how did a swirling mix of chemicals develop into a living thing? How did chemistry become... biology?

Nobody is totally sure, but we believe it required two things: **CHEMICALS** and **ENERGY**.

Perhaps the thermal energy within an undersea vent was enough to cause a change... ...at the molecular level.

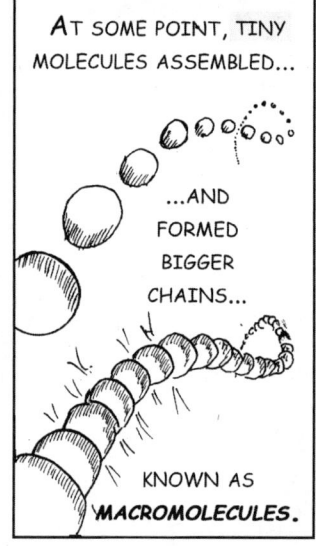

At some point, tiny molecules assembled... ...and formed bigger chains... known as **MACROMOLECULES**.

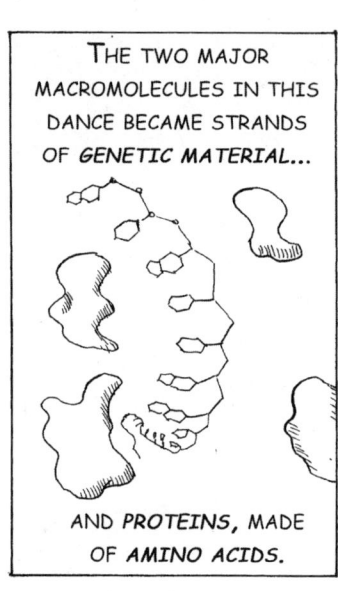

The two major macromolecules in this dance became strands of **GENETIC MATERIAL**... and **PROTEINS**, made of **AMINO ACIDS**.

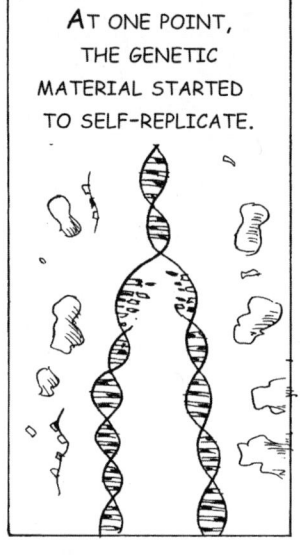

At one point, the genetic material started to self-replicate.

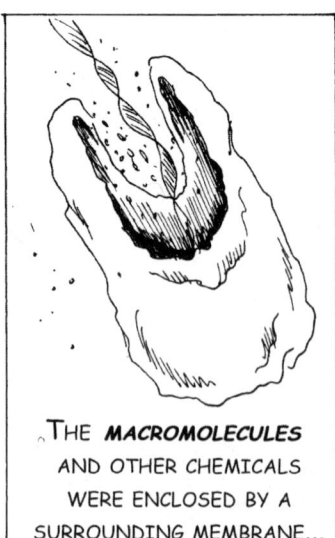

The **MACROMOLECULES** and other chemicals were enclosed by a surrounding membrane...

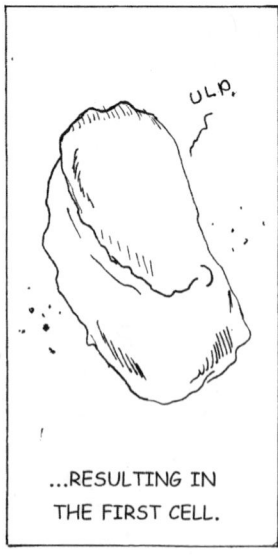

ULP.

...resulting in the first cell.

We think life originated in undersea vents because of their ability to support microbial life, their access to energy, and their rich supply of chemicals.

Ha! Call me when you have some **REAL** proof!

DR. BIRDLEY INVESTIGATES: Living Fossils

Origin of Life

Geologic Time

Objectives

1. To describe a plausible scenario for where, when, and how life originated.
2. To introduce ocean floor hydrothermal vents.

Synopsis

Dr. Birdley explains a theory about the origin of life to the skeptical Dean Owelle.

Main Ideas

1. At some point in Earth's history, non-living molecules became organized into living systems.
2. Earth was still a harsh place 3.8 billion years ago.
3. Undersea vents may have been the sites at which life first originated because of their distance from Earth's surface, proximity to an energy source, plethora of organic compounds, and capacity to support microbial life.
4. Organic molecules combined to make larger molecules, such as DNA and proteins.
5. This mix of DNA and proteins ultimately became enclosed by a membrane, forming the first cell.
6. The first living things were single-celled.

Vocabulary

polymers	amino acids	organic compounds
replicate	molecular	hydrothermal vents
cell	membrane	genetic material
chemistry	biology	macromolecules

Characters

Dr. Birdley, Dean Owelle, and the first microbe.

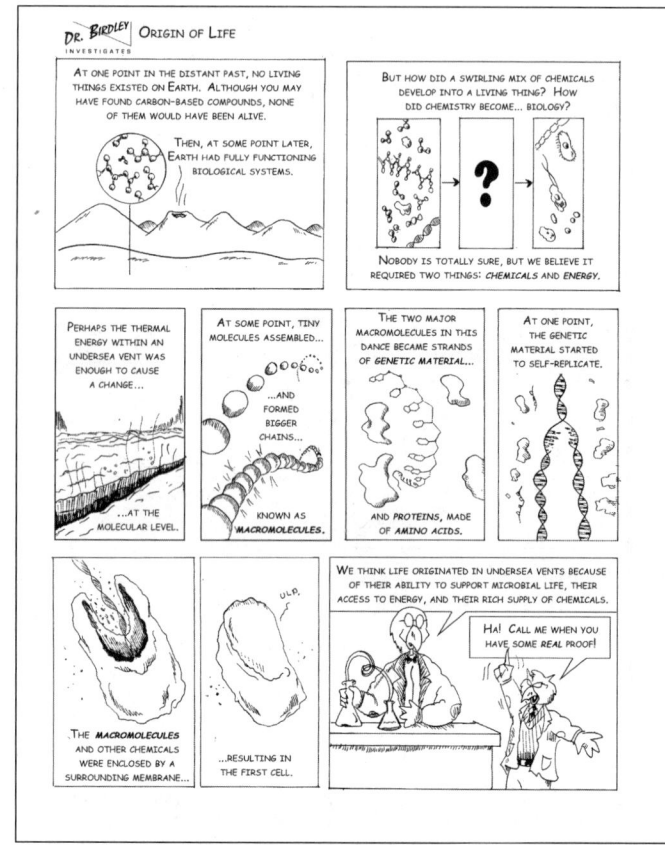

Questions for Discussion

Before Reading:

1. How long do you think living things have been around?
2. What is the difference between chemistry and biology?
3. How are the two subjects similar?
4. What do all living things need to survive?
5. What are all living things made of?

After Reading:

1. Do you agree with this idea? Why or why not?
2. What events beneath the earth may be related to these hydrothermal vents?

Living Fossils

Microbes & Geology

Objectives

1. To present examples of present-day organisms that are thought to be similar to the first organisms on Earth.
2. To illustrate how microbes can shape the interior of a cave by producing acidic chemicals.

Synopsis

Dr. Birdley and Gina explore the Cueve de Villa Luz (Cave of the Lighted House) in Southern Mexico, where they find an unusual microbial habitat.

Main Ideas

1. Snottites (drippy, gooey, acidic projections that contain millions of microbes) hang from the ceilings of the Cueve de Villa Luz.
2. The extremophilic microbes within snottites are thought to be similar to the first living things on Earth.
3. These microbes metabolize hydrogen sulfide, which is poisonous to humans.
4. The microbes produce sulfuric acid, which carved through the limestone in the cave, producing passages.

Vocabulary

limestone	hydrogen sulfide	sulfuric acid
archaea	extremophiles	limestone
snottites	microbes	acidic

Characters

Dr. Birdley, Gina

Teacher's Note

The microbes that produce and consume the sulfur-based compounds contribute to the sulfur cycle. The visual exercise covers the sulfur cycle in broader detail.

Questions for Discussion

Before Reading:

1. What do you think were the first living things on Earth?
2. What does it mean for a substance to be acidic?
3. How do you think caves form?

After Reading:

1. Why did Dr. Birdley go rushing out of the cave?
2. If these types of microbes were similar to the first living things on Earth, what does this tell you about the likely conditions on Earth back then?

ORIGIN OF LIFE

NAME: _____

CLASS: _____ DATE: _____

STUDY QUESTIONS

Directions: Answer the following questions to the best of your ability.

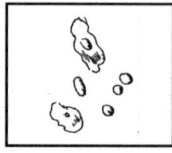

1. What two resources on Earth were required for life to emerge? What is one theory for how these resources were provided?
_____
_____

2. What are some qualities of undersea hydrothermal vents that make them likely place for the origin of life?
_____
_____

3. Describe the suggested process by which the macromolecules were produced. Include all the major steps mentioned in the comic.
_____
_____

4. Describe the suggested events that caused the first living cell to form.
_____
_____

5. Why is this idea for how life originated still uncertain?
_____
_____

 LIVING FOSSILS

NAME: _____

CLASS: _____ DATE: _____

 STUDY QUESTIONS

Directions: Answer the following questions to the best of your ability.

1. Why are the microbes within the snottites referred to as living fossils?
   _____
   _____

2. What are snottites? What do you find inside them?
   _____
   _____

3. Why is it dangerous to enter the cave without protective clothing?
   _____
   _____

4. Explain why this environment is good for these microbes but harmful for humans.
   _____
   _____

5. How do microbes help to carve out the tunnels of the caves?
   _____
   _____

Copyright ©2009 by Incentive Publications, Inc., Nashville, TN

Dr. Birdley Teaches Science – Mysteries of the Earth

39

DR. BIRDLEY INVESTIGATES

ORIGIN OF LIFE

NAME: _____

CLASS: _____ DATE: _____

BACKGROUND: DID LIFE ORIGINATE IN A VENT?

At some point in Earth's history, living systems emerged from non-living organic molecules. We do not know for certain where this occurred, how it occurred, or how many times it had to happen before life on Earth could sustain itself. Even in labs, nobody has been able to create conditions under which life arises from non-life.

So where did life first originate? One plausible theory is that the emergence of life took place in the hydrothermal vents—fissures in the ocean floor that existed when the Earth was young, and that continue to serve as habitats for microbial communities today. It is through these fissures that magma, ash, and rocky debris flow, temperatures are extremely hot. Yet, despite these harsh conditions, life continues to persist in these habitats.

PERHAPS THE THERMAL ENERGY WITHIN AN UNDERSEA VENT WAS ENOUGH TO CAUSE A CHANGE...

...AT THE MOLECULAR LEVEL.

Several characteristics make vents likely candidates for the "origin of life." They were far away from Earth's surface, which was likely to have been too harsh to support life. The vents also contained a rich mix of organic chemicals containing nitrogen, sulfur, oxygen, and carbon—the necessary elements that all living things are composed of. And the thermal energy arising from within the Earth would have provided the energy needed for life to emerge and sustain itself.

Support for the vent hypothesis also comes from the microbial communities that exist within the vents today, which tolerate the extreme temperatures and consume the available chemicals there. Scientists examine these microbes to learn more about what the first living things on Earth may have been like.

Directions: Answer the following questions to the best of your ability.

1. Why are hydrothermal vent likely candidates for the sites at which life first originated?

_____

_____

_____

2. What types of living things exist in hydrothermal vents? What qualities would they need to have in order to survive and thrive in this habitat?

_____

_____

_____

Copyright ©2009 by Incentive Publications, Inc., Nashville, TN

Dr. Birdley Teaches Science – *Mysteries of the Earth*

The Sulfur Cycle

NAME: _____

CLASS: _____ DATE: _____

Label the diagram of the sulfur cycle with the correct terms! Use the word bank.

WORD BANK

ACID	LIMESTONE
RAIN	HYDROGEN SULFIDE (2)
SULFURIC (2)	SULFIDE
SULFUR DIOXIDE (2)	VOLCANOES
PRODUCE	MICROBES
DEAD MATTER	SALTS

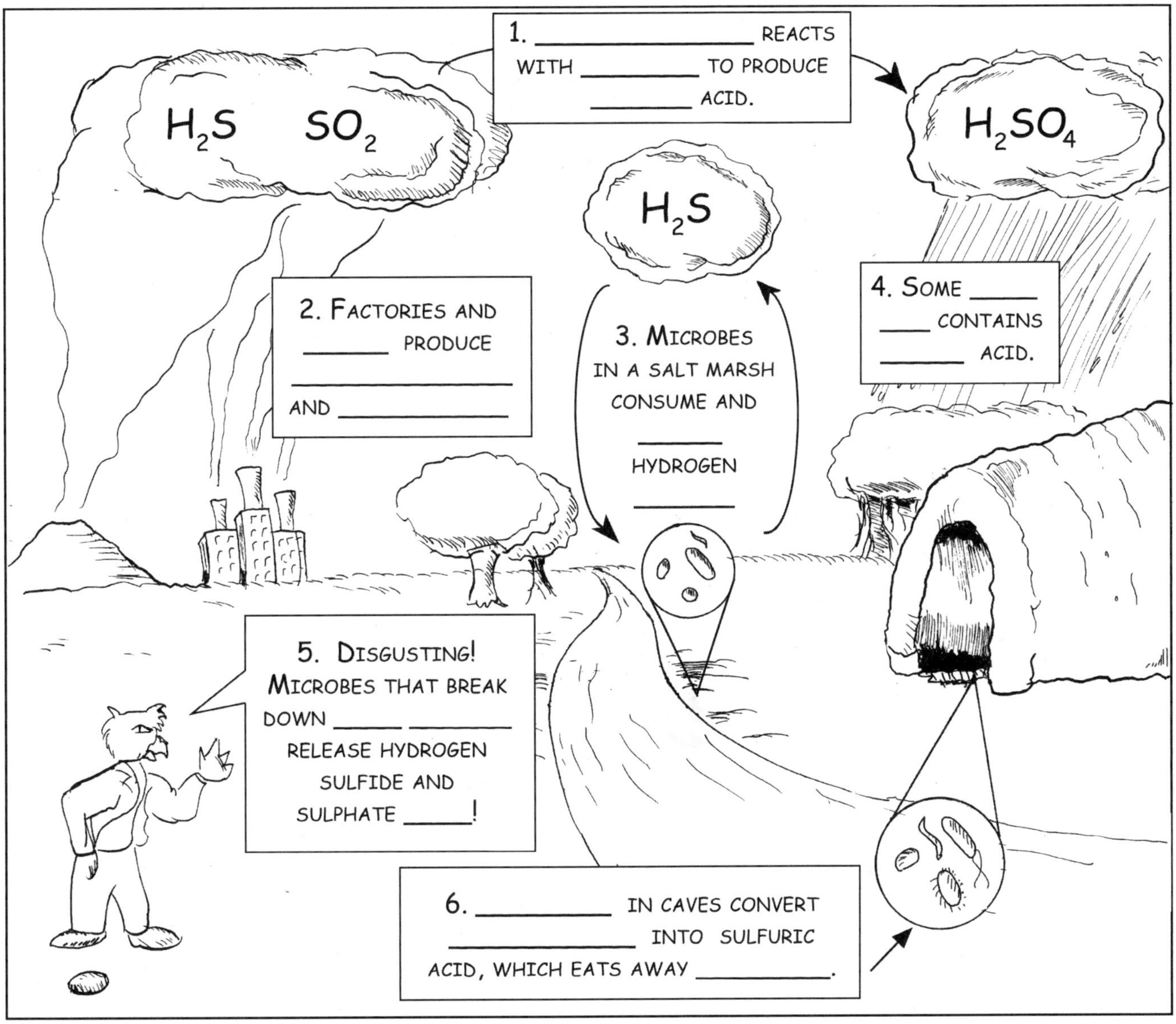

1. _____ REACTS WITH _____ TO PRODUCE _____ ACID.

2. FACTORIES AND _____ PRODUCE _____ AND _____.

3. MICROBES IN A SALT MARSH CONSUME AND _____ HYDROGEN _____.

4. SOME _____ _____ CONTAINS _____ ACID.

5. DISGUSTING! MICROBES THAT BREAK DOWN _____ _____ RELEASE HYDROGEN SULFIDE AND SULPHATE _____!

6. _____ IN CAVES CONVERT _____ INTO SULFURIC ACID, WHICH EATS AWAY _____.

Unit 4: Microbial Planet

Contents

Background: Biogeochemical Cycles 43

Source Cartoon: Biogeochemical Cycles 44

Source Cartoon: Three Domains 45

Cartoon Profiles 46

Study Questions 48

Background 50

Visual Exercise 51

Quiz 52

A MICROBIAL PLANET

NAME: _____

CLASS: _____ DATE: _____

BACKGROUND: BIOGEOCHEMICAL CYCLES

Microbes are usually thought of as the disease-causing agents of dirty water and foul air. However, they are only a minority. Most bacteria are not harmful, and many of them assist in the **biogeochemical cycles**—continuous processes by which key elements are recycled and moved through the biosphere. Carbon, oxygen, sulfur, nitrogen, and phosphorous each go through their own unique biogeochemical cycle, thanks in large part to the microbes.

Carl Woese's "tree of life" illustrates the incredible genetic diversity among **bacteria** and **archaea**. Genetically, bacteria are more different from archaea than a human is from a mushroom. This genetic diversity results in great metabolic diversity—there are tremendous

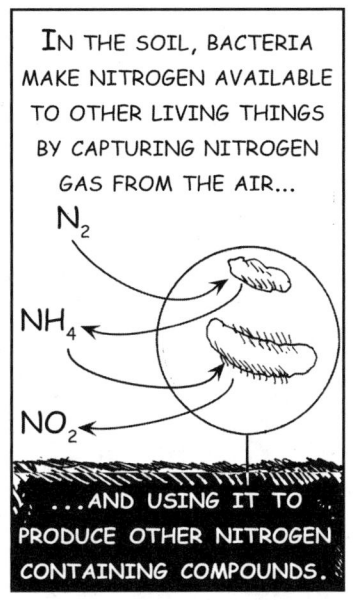

IN THE SOIL, BACTERIA MAKE NITROGEN AVAILABLE TO OTHER LIVING THINGS BY CAPTURING NITROGEN GAS FROM THE AIR...

N_2
NH_4
NO_2

...AND USING IT TO PRODUCE OTHER NITROGEN CONTAINING COMPOUNDS.

differences in the compounds that microbes eat, breathe, and produce. Some microbes are able to recycle the carbon in the soil and send it back into the atmosphere. Others were able to harness nitrogen gas (N_2) from the air and convert it into compounds such as ammonium (NH_4), which other life forms could readily use.

By setting the biogeochemical cycles in motion, microbes in the Precambrian time set the stage for later life. The carbon, nitrogen, sulfur, and oxygen cycles are processes that keep these critical elements moving through both nonliving and living parts of the biosphere, making and breaking bonds with different compounds. All of Earth's ecosystems need this critical recycling of resources in order for them to sustain themselves.

Directions: Answer the following questions to the best of your ability.

1. Explain how microbes set the stage for later life during Precambrian Time.

_____
_____
_____
_____

2. What are biogeochemical cycles? List three elements that go through a biogeochemical cycle.

_____
_____
_____
_____

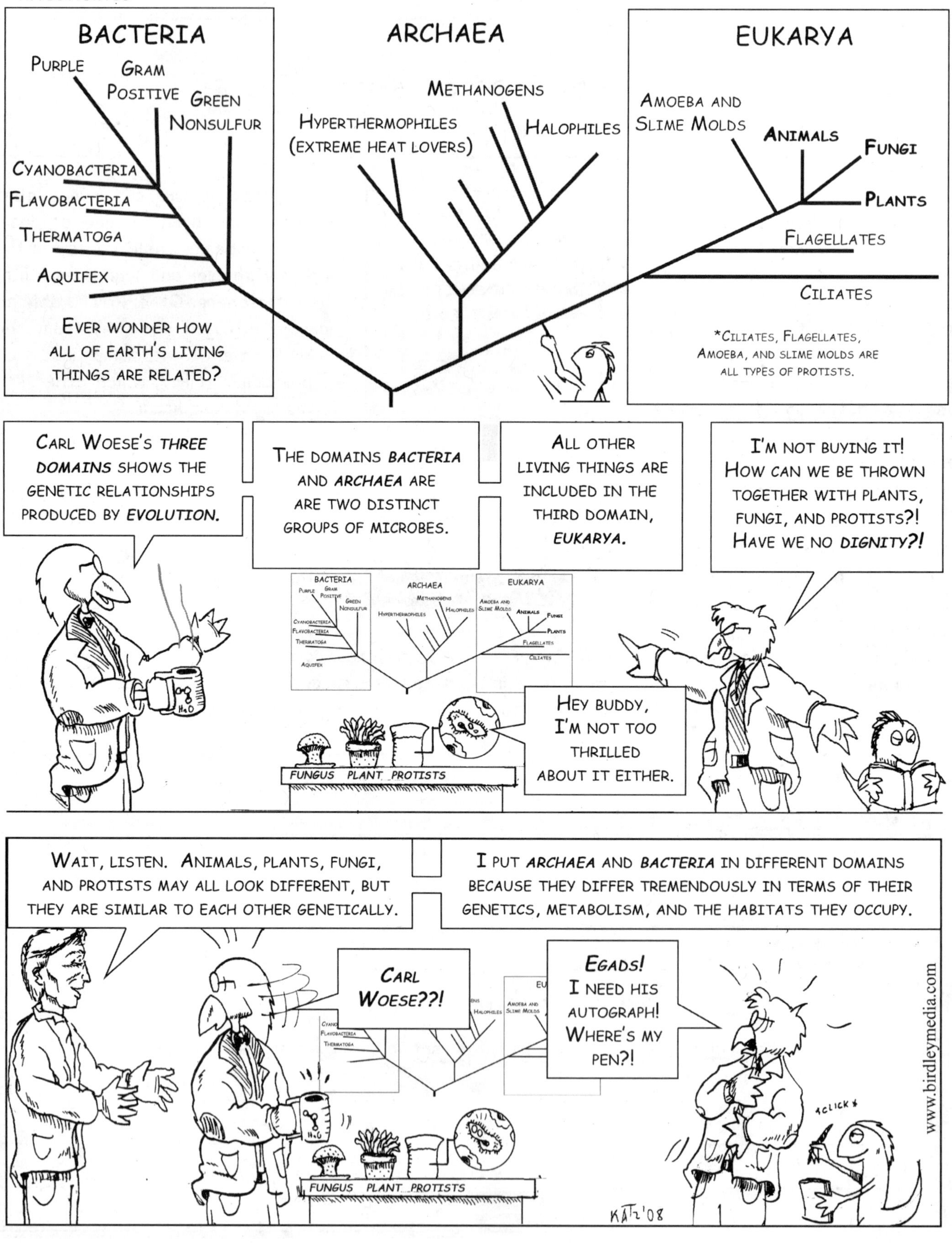

NAME:_____

CLASS:_____ DATE:_____

 **BIOGEOCHEMICAL CYCLES**

"You're planning a lesson on Precambrian time?! Why teach about such an uneventful time period?!"

"Actually, the Precambrian was a critical time for the planet."

"It was during this time that microbes started recycling the key elements, like oxygen, carbon, nitrogen, and sulfur."

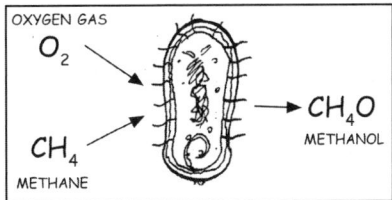

"Among the millions of different microbes, each one consumed specific compounds and produced important new substances."

"By doing this, the microbes set up the *BIOGEOCHEMICAL CYCLES*, which still sustain the world's ecosystems today."

"Let's look at some roles that microbes play in these cycles."

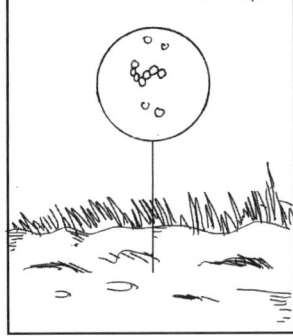

"In the oxygen-free depths of salt marshes and swamps, microbes convert carbon dioxide (CO_2) into methane (CH_4)."

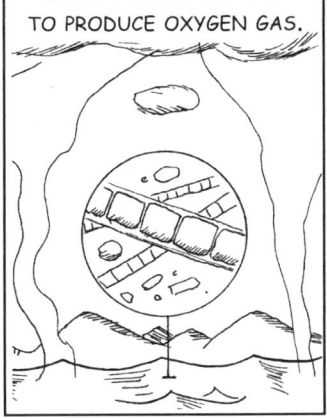

"On the ocean surface, microbes use water, carbon dioxide, and energy from the sun to produce oxygen gas."

"And on the ocean floor, microbes in undersea vents use *GEOTHERMAL ENERGY* to convert sulfate into hydrogen sulfide."

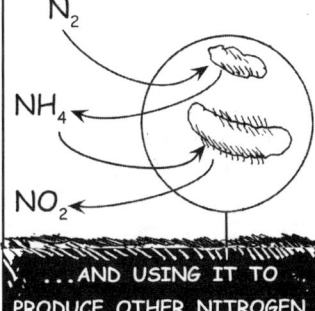

"In the soil, bacteria make nitrogen available to other living things by capturing nitrogen gas from the air...

N_2
NH_4
NO_2

...and using it to produce other nitrogen containing compounds."

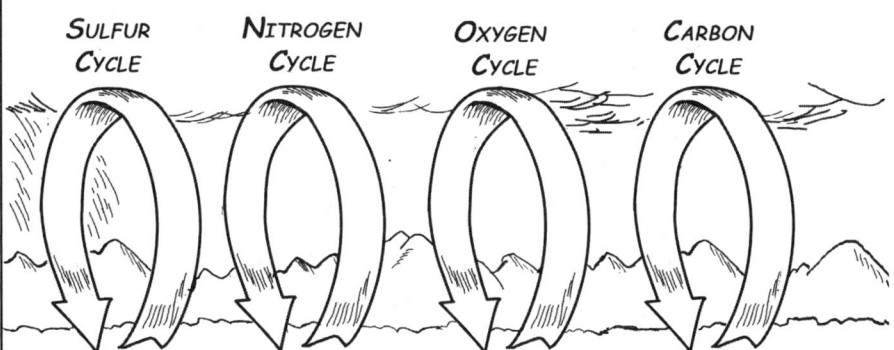

"Because key elements occur in a variety of compounds, many different microbes are needed to recycle these elements in a variety of ways."

SULFUR CYCLE — NITROGEN CYCLE — OXYGEN CYCLE — CARBON CYCLE

"Together, they send matter through both living and nonliving parts of the ecosystem, resulting in the biogeochemical cycles that occur all over the Earth."

"Good gravy! When did these microbes start running the planet?!"

"About 3.8 billion years ago."

THE THREE DOMAINS

The Biosphere

Objectives

1. To introduce how all living things are classified under the three domain scheme.
2. To present Earth's extensive microbial diversity as the rationale for the three domains.

Synopsis

Dr. Birdley explains the three domains to the skeptical Dean Owelle, who ends up dropping his argument for the moment when the creator of the theory, Dr. Carl Woese, pays them a visit.

Main Ideas

1. The three domains is a classification scheme for all living things based on DNA sequence data from a great many living things.
2. In this scheme, two domains, archaea and bacteria, are microbial, while eukarya is comprised of animals, plants, protists, and fungi.
3. Animals, plants, protists and fungi are more similar to each other genetically than archaea and eukarya.
4. Eukarya and archea demonstrate tremendous diversity in terms of genetics, metabolism, and habitat.

Vocabulary

domain	evolution	genetics
metabolism	phylogenetic	microbe
eukarya	archaea	bacteria

Characters

Carl Woese, Dr. Birdley, Dean Owelle, Norman, and a sarcastic protist.

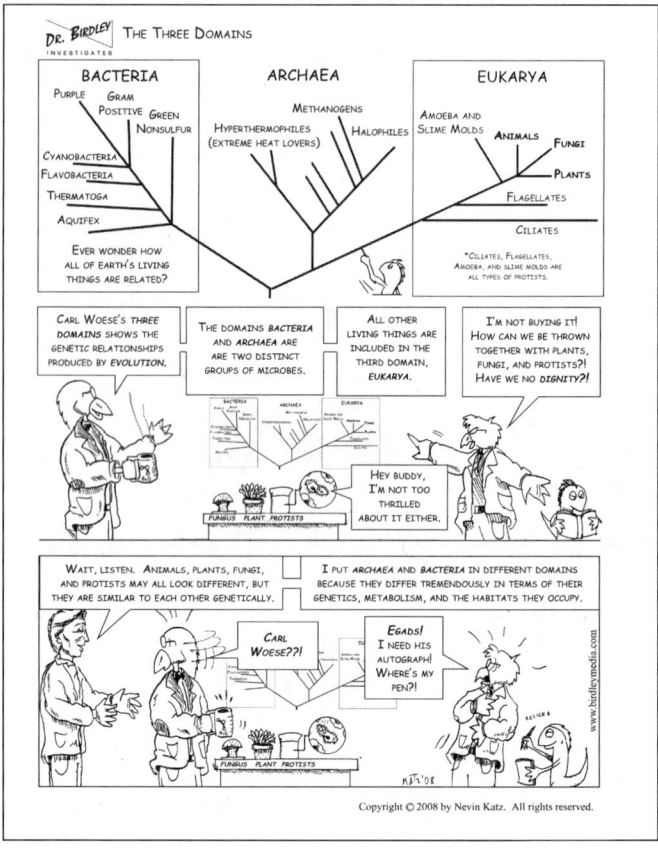

Questions for Discussion

Before Reading:

1. Make a list of all the major categories of living things on Earth.
2. Give an example of two animals or plants that you think are genetically similar. Why do you think they are similar?
3. What do you already know about DNA? Bacteria?

After Reading:

1. Is it possible to have more specific phylogenetic trees within the three domains tree? Why or why not?
2. Would you say that the three domains tree is general or specific?

BIOGEOCHEMICAL CYCLES

The Biosphere

Objectives

1. To relate the Precambrian Time to the biogeochemical cycles.
2. To explain the importance of microbes in the biogeochemical cycles.
3. To introduce typical microbial habitats.
4. To provide examples of how microbes recycle elements.

Synopsis

In response to Owelle's skeptical remarks, Dr. Birdley explains to him the significance of the Precambrian and biogeochemical cycles.

Main Ideas

1. Microbes set up biogeochemical cycles during Precambrian Time.
2. In biogeochemical cycles, a key element will travel through living and nonliving parts of the ecosystem, incorporating itself into different types of substances.
3. Bacteria act as essential agents of change in these cycles, chemically transforming compounds and producing new substances.
4. Because of their metabolic diversity, bacteria are present at multiple stages of the biogeochemical cycles.

Vocabulary

biosphere	biogeochemical	cycle
metabolism	microbes	hydrogen
sulfur	carbon	nitrogen
Precambrian	oxygen	methane
sulfate	hydrogen sulfide	ammonium

Characters

Dean Owelle, Dr. Birdley, Norman

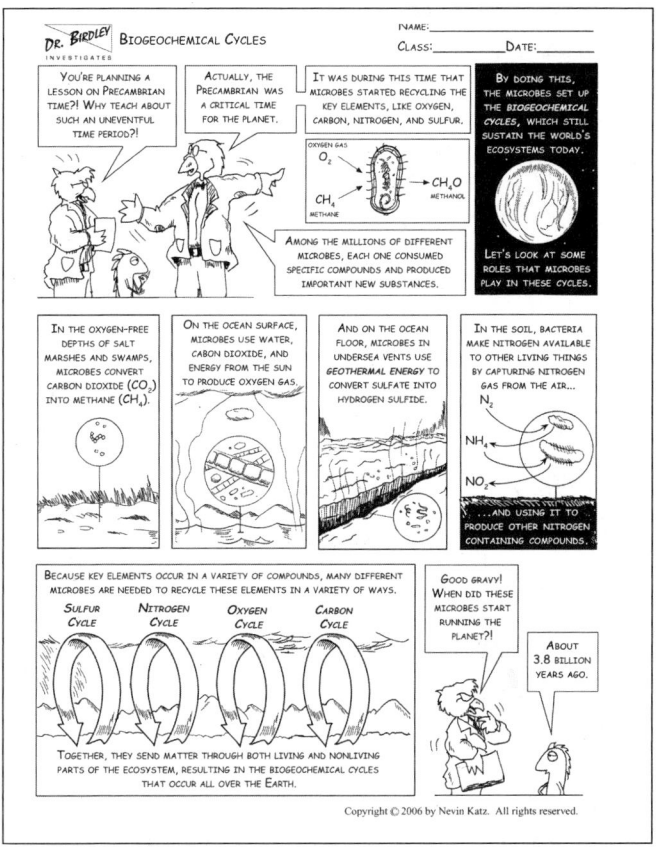

Questions for Discussion

Before Reading:

1. What do you already know about the matter cycles?
2. What are the various types of compounds that an oxygen atom could be a part of?
3. What are the meanings of the prefixes *bio-* and *geo-*?

After Reading:

1. How are microbes like chemical factories?
2. Which habitats would you be most interested in exploring?
3. So why was the Precambrian important?
4. Why are microbes important?

THE THREE DOMAINS

NAME: _____

CLASS: _____ DATE: _____

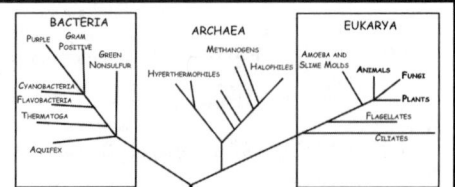

STUDY QUESTIONS

Directions: Answer the following questions to the best of your ability.

1. What is the diagram in the top panel? What does it illustrate?
   _____
   _____

2. What does the three domains scheme imply about the diversity of microbes relative to animals?
   _____
   _____

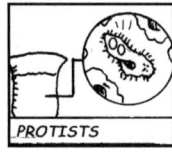

3. Why are animals, plants, fungi, and protists together?
   _____
   _____

4. Name several ways in which microbes are diverse.
   _____
   _____

5. What types of data was needed to be collected to construct the three domains tree?
   _____
   _____

 **Biogeochemical Cycles**

Name: _____

Class: _____ Date: _____

Study Questions

Directions: Answer the following questions to the best of your ability.

1. What is it about a microbe that allows it to play such a key role in biogeochemical cycles?

   _____

   _____

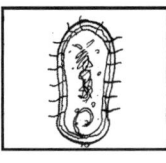
2. What two substances does the first microbe at the top of the comic use? Where do you see these substances being produced in the middle panels?

   _____

   _____

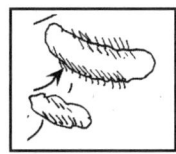
3. No animals can obtain nitrogen directly from the air. How do microbes make nitrogen accessible to other living things?

   _____

   _____

4. What is a biogeochemical cycle?

   _____

   _____

5. Why is it important for the world's population of microbes to be extremely diverse?

   _____

   _____

The Carbon Cycle

Name: _____
Class: _____ Date: _____

Label the diagram of the carbon cycle with the correct terms! Use the word bank.

"Pass the carbon."

WORD BANK

AIR	RELEASED
PHOTOSYNTHETIC	TREES
WASTE	CARBON DIOXIDE
ENERGY	MICROBES
RESPIRATION	FOSSIL FUELS (2)
PHOTOSYNTHESIS	DECOMPOSITION

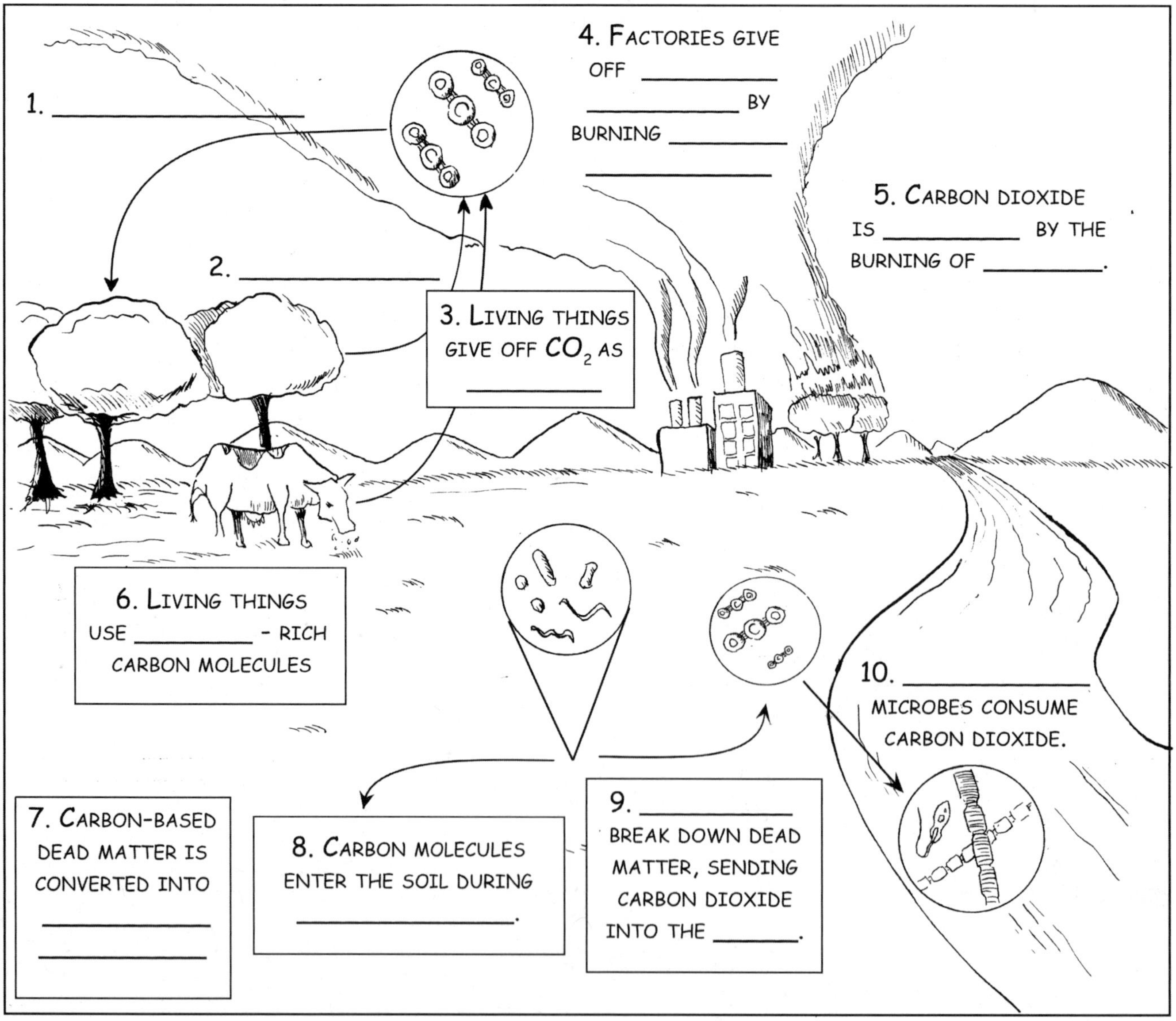

1. _____

2. _____

3. Living things give off CO_2 as _____

4. Factories give off _____ by burning _____

5. Carbon dioxide is _____ by the burning of _____.

6. Living things use _____ - rich carbon molecules

7. Carbon-based dead matter is converted into _____ _____

8. Carbon molecules enter the soil during _____.

9. _____ break down dead matter, sending carbon dioxide into the _____.

10. _____ microbes consume carbon dioxide.

Name:_____ Class:_____ Date:_____

Unit 3 and 4 Quiz: Microbes and the Matter Cycles

Directions: This quiz tests your knowledge of the chapter's cartoon, background article, and visual exercises. Answer the following questions to the best of your ability.

1. The transformation of nitrogen gas into useful compounds is made possible by chemical reactions that take place within
   a) humans
   b) bacteria
   c) volcanoes
   d) clouds

2. Undersea vents are a likely spot for the origin of life, partially because of high amounts of
   a) geothermal energy
   b) mineral fluids
   c) sedimentary rocks

3. In a biogeochemical cycle, matter travels through only the living parts of the ecosystem.
   a) true
   b) false

4. The three domains emphasize the diversity of
   a) mammals
   b) plants
   c) fungi
   d) micro-organisms
   e) reptiles

5. Carbon dioxide is released into the air as a result of a natural process known as
   a) photosynthesis
   b) erosion
   c) cementation
   d) decomposition

6. Microbes may accelerate the breakdown of limestone cave walls by producing
   a) carbon dioxide
   b) methane
   c) sulfuric acid
   d) formaldehyde

7. Which of the following is most directly caused by the sulfur dioxide in factory emissions reacting with the water in clouds?
   a) smog
   b) acid rain
   c) deterioration of the ozone layer
   d) global warming

8. Carbon is stored in fossil fuels.
   a) true
   b) false

7. Give an example of the role a micro-organism may play in a biogeochemical cycle.

_____

_____

_____

_____

Unit 5: Introducing Rocks

Contents

Vocabulary Build-up 53

Source Cartoon: Types of Rocks 54

Source Cartoon: Transforming Rocks 55

Cartoon Profiles 56

Study Questions 58

Background 60

Visual Exercise 61

Quiz 62

 ROCKS

NAME: _____

CLASS: _____ DATE: _____

 VOCABULARY BUILD-UP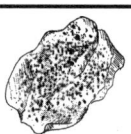

Part A Fill in each blank with the correct word from the word bank.

1. **Igneous rock** - a rock formed from _____ or _____ that has cooled off.

2. **Sedimentary rock** - a rock formed from _____ that have been _____ together.

3. **Metamorphic rock** - rock that has been _____ due to _____ and _____

WORD BANK
sediments   lava   transformed   compacted   pressure   magma   heat

Part B Convey the meanings of the three rock types above by filling in the word balloons.

TRANSFORMING ROCKS

NAME:_____
CLASS:_____ DATE:_____

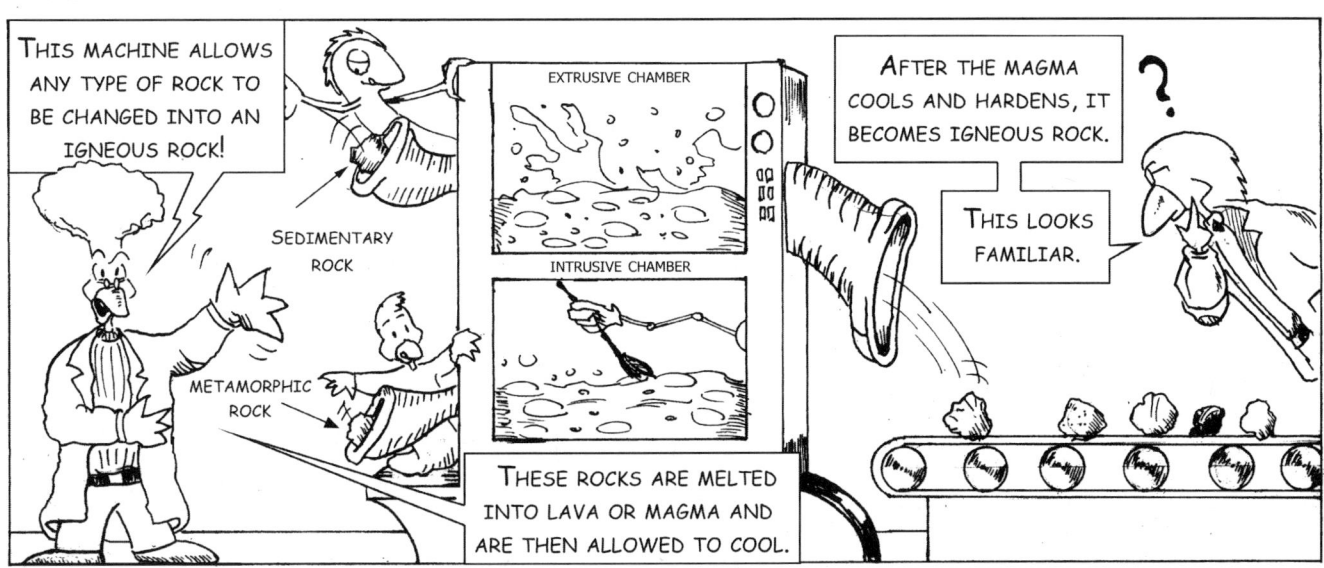

Types of Rocks

Geology

Objectives

1. To describe central characteristics of the three major types of rocks.
2. To explain how each major type of rock forms.
3. To provide key examples of each type of rock.

Synopsis

Birdley, Jaykes, and Gina provide an overview of the major types of rocks, while standing upon examples of each of them. Intrusive igneous rocks spring up from within the Earth. Norman busily sprinkles rock salt, a type of sedimentary rock. A statue of Owelle stands that is made of marble, which is a metamorphic rock.

Main Ideas

1. Igneous rocks form from molten rock, which is either magma or lava. Examples include scoria, basalt, pumice, andesite, and granite.
2. Sediments may come from erosion and weathering, biological sources, or substances precipitating out of solution. Examples include sand, silt, gravel, rock salt, limestone, sandstone, clay, chalk.
3. Rocks may transform into metamorphic rock under tremendous heat and pressure. Examples include gneiss, marble, schist, quartzite, skarn, and slate.

Vocabulary

sedimentary	igneous	metamorphic
magma	weathering	mineral fluid
temperature	pressure	compress

Characters

Dr. Birdley, Gina, Jaykes, and Norman

Questions for Discussion

Before Reading:

1. What are some differences that might show up in two or more rocks that you find?
2. What are some properties of rocks that can be observed and described?
3. How are all rocks similar?

After Reading:

1. Describe the different ways in which rocks can form.
2. What are some examples of rocks that you recognize in the picture?
3. Why do you think the rocks are coming out of the Earth?

Copyright ©2009 by Incentive Publications, Inc., Nashville, TN

Dr. Birdley Teaches Science – Mysteries of the Earth

Transforming Rocks

Rock Cycle

Objectives

1. To highlight the major processes in the rock cycle.
2. To explain the rock cycle in terms of what processes are needed to produce a particular type of rock.

Synopsis

Professor Lark invents a set of machines that transform rocks from one type to another. Birdley points out that all these processes take place within the rock cycle, which occurs naturally.

Main Ideas

1. Given the right conditions, any rock could become an igneous, metamorphic, or sedimentary rock.
2. Heat and pressure are required to transform a rock into a metamorphic rock.
3. Rocks are broken into sediments, which are then forced together by pressure.
4. In some cases, mineral fluids will act as glue, causing the sediments to stick together.
5. Any rock could become an igneous rock if it is melted into magma or lava and allowed time to harden and cool.
6. Rocks that form above ground are intrusive, and rocks that form below ground are extrusive.

Vocabulary

igneous	sedimentary	metamorphic
heat	pressure	molten rock
lava	magma	sediments
compression	cementation	weathering

Characters

Professor Lark, Norman, Baby Birdley, Dr. Birdley, Jaykes, Gina, Don

Questions for Discussion

Before Reading:

1. How do you think one rock changes into another rock?
2. How might wind and rain affect a rock?
3. What events or forces may cause a rock to transform?

After Reading:

1. How do these machines simulate the rock cycle?
2. Do you consider this set of machines to be useful? Why or why not?
3. Which machine do you find the most complex? Why?

 ROCK TYPES

NAME: _____
CLASS: _____ DATE: _____

STUDY QUESTIONS

Directions: Answer the following questions to the best of your ability.

 1. Describe the process that results in the formation of igneous rocks.
_____
_____

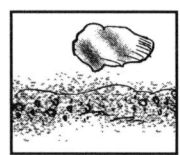 2. What are two ways sediments could be created?
_____
_____

 3. How are sediments turn into sedimentary rock?
_____
_____

 4. What do all metamorphic rocks have in common?
_____
_____

 5. Give an example of how a process (weathering, compaction, cooling, heating) could contribute to the formation of a particular rock type.
_____
_____

DR. BIRDLEY INVESTIGATES

TRANSFORMING ROCKS

NAME: _____

CLASS: _____ DATE: _____

STUDY QUESTIONS

Directions: Answer the following questions to the best of your ability.

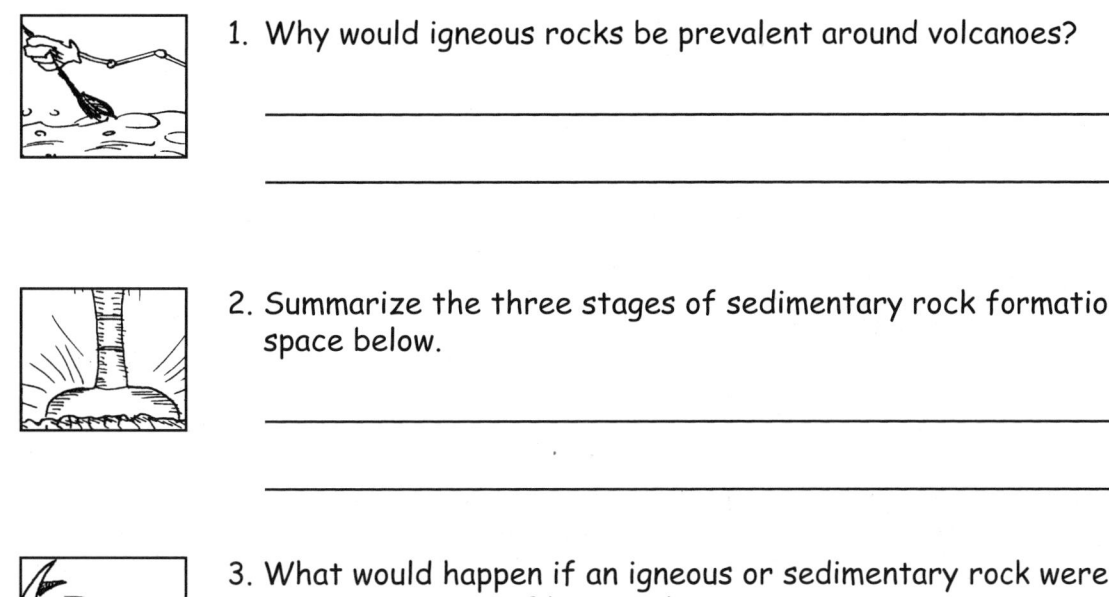

1. Why would igneous rocks be prevalent around volcanoes?
_____
_____

2. Summarize the three stages of sedimentary rock formation in the space below.
_____
_____

3. What would happen if an igneous or sedimentary rock were exposed to intense amounts of heat and pressure?
_____
_____

4. Give an example of how erosion and weathering in Lark's second machine may differ from how it may occur in the natural rock cycle.
_____
_____

5. Is it possible for an igneous rock to become sedimentary, and then become metamorphic? Explain.
_____
_____

INTRODUCING ROCKS

NAME: _____

CLASS: _____ DATE: _____

BACKGROUND: TYPES OF ROCKS

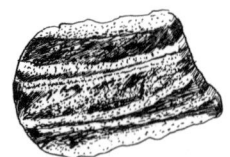 

Directions: Read the comic and text. Then, answer the questions to the best of your ability.

A rock may be igneous, sedimentary, and metamorphic. Rocks change from one of these forms to another in a continual process known as the rock cycle, which involves weathering, compaction, heating, cooling, and other natural processes. Igneous rocks are formed from molten rock that has cooled. Sedimentary rocks are formed from sediments that have been compacted together. Metamorphic rocks arise from rocks that have been transformed due to tremendous heat and pressure from deep within the Earth. The pressure is usually due to collisions between tectonic plates.

Rocks can be identified by examining properties, such as texture, lustre, hardness, and the patterns of mineral crystals on their surface. On foliated metamorphic rocks, for example, alternating bands of mineral crystals are visible. Nonfoliated metamorphic rocks, such as marble and quartzite, are not banded. The rock in the comic is gneiss, a foliated metamorphic rock that was formed from granite. During its formation, the intense heat caused minerals to melt and recrystallize, arranging themselves into bands.

Directions: Answer the following questions to the best of your ability.

1. What is the difference between a foliated and a non-foliated metamorphic rock?

_____

_____

2. Why might metamorphic rocks form near tectonic plate boundaries?

_____

_____

THE ROCK CYCLE

NAME: _____

CLASS: _____ DATE: _____

LABEL EACH PROCESS OF THE ROCK CYCLE USING THE WORD BANK! YOU WILL USE MOST WORDS MORE THAN ONCE.

WORD BANK

COOLING

MELTING

HEAT AND PRESSURE

WEATHERING AND EROSION

COMPACTION AND CEMENTATION

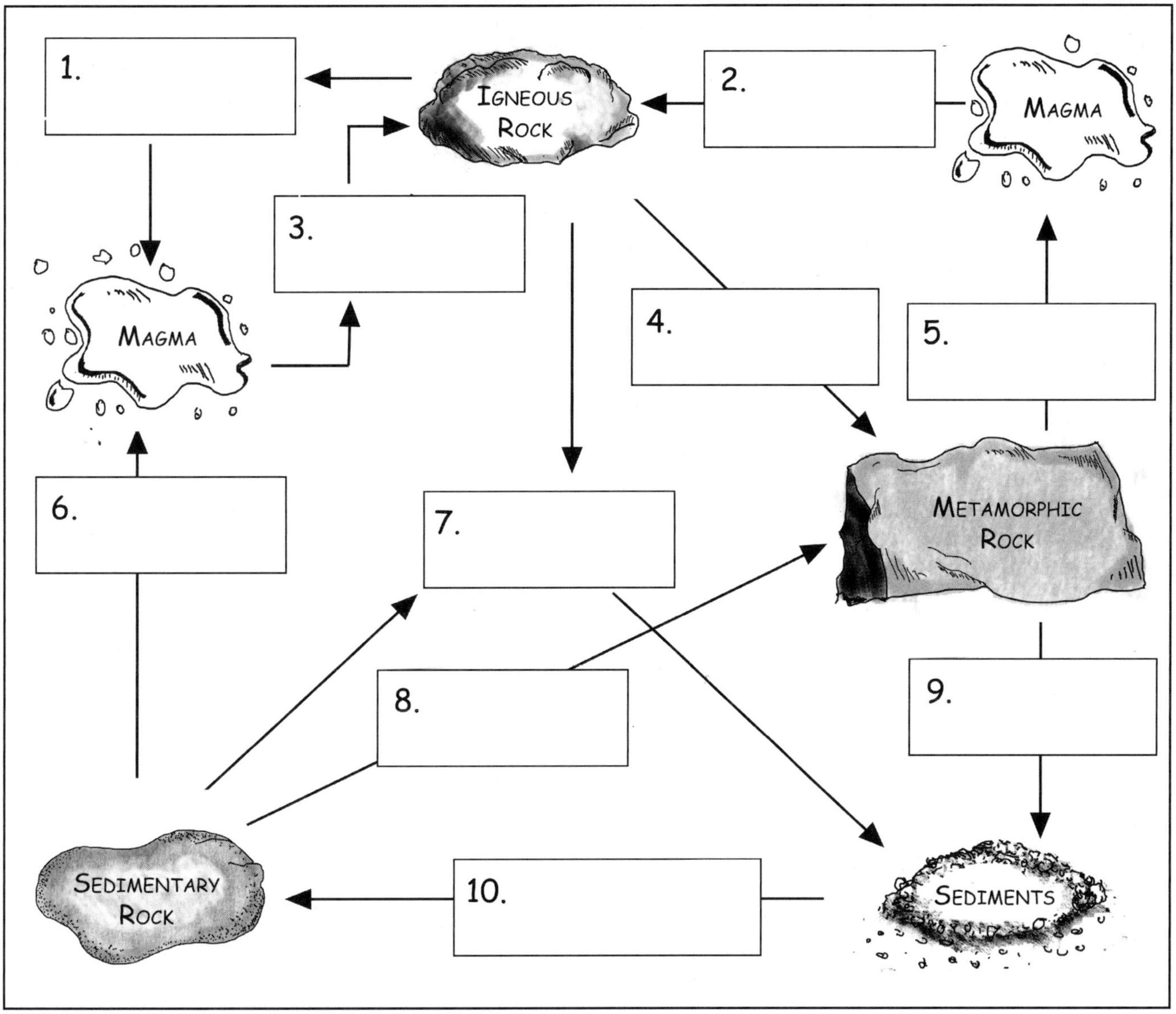

Copyright ©2009 by Incentive Publications, Inc., Nashville, TN

Dr. Birdley Teaches Science – Mysteries of the Earth

Name: _____ Class: _____ Date: _____

Unit 5 Quiz: Classifying Rocks and the Rock Cycle

Directions: Answer each question to the best of your ability.

Part A. Identify the type of rock that each geological feature is made of. Possible answers include *igneous*, *sedimentary*, or *metamorphic*.

THE MOUNTAINS OF CHURCHILL IN CANADA ARE MADE OF QUARTZITE, WHICH WAS FORMED FROM SANDSTONE UNDER INTENSE HEAT AND PRESSURE.

THE SALT CELLAR TOR IN ENGLAND IS FORMED FROM LAYERS OF COMPRESSED SHALE AND SANDSTONE THAT WERE WEATHERED UNDERGROUND.

THE HEXAGONAL COLUMNS OF THE GIANT'S CAUSEWAY IN IRELAND ARE FORMED FROM A LAKE OF BASALT LAVA THAT COOLED SLOWLY OVER TIME.

1. _____   2. _____   3. _____

Part B. Answer the following questions.

4. Which of the following processes contributes most to the formation of igneous rock?
   - (a) compaction of pebbles
   - (b) cooling of lava or magma
   - (c) weathering and erosion
   - (d) transformation under pressure

5. The process by which mineral fluids help sediments stick together is called
   - (a) compaction
   - (b) weathering
   - (c) cementation
   - (d) deposition

6. A geologist informs you that a metamorphic rock known as slate may have been at some point a sedimentary rock known as shale. Explain how this transformation could have happened.

_____

_____

_____

_____

Unit 5: Sedimentary Rocks

Contents

Source Cartoon: Owelle's Rock 64

Source Cartoon: White Sands 65

Cartoon Profiles 66

Study Questions 68

Source Cartoon: A Lesson from a Snail 70

Cartoon Profile 71

Study Questions 72

Background 73

Visual Exercise 74

Mini-Comic: Cliffs of Dover 75

Graphic Organizer 76

Owelle's Rock

Name:_____
Class:_____ Date:_____

White Sands

Name:_____
Class:_____ Date:_____

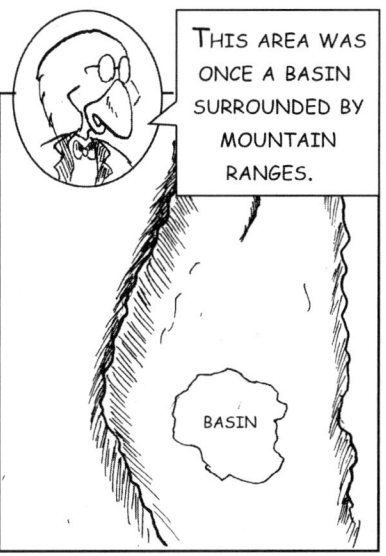

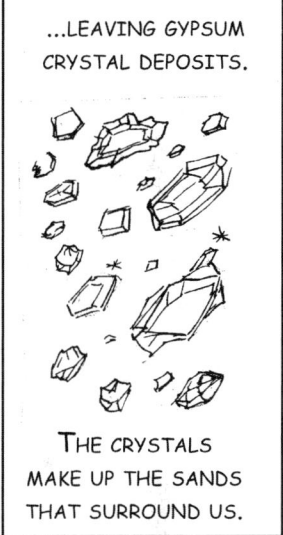

Owelle's Rock

Sedimentary Rocks

Objectives

1. To illustrate the basic processes in the formation of a detrital sedimentary rock.
2. To illustrate the potential effects of weathering and erosion on a rock.

Synopsis

Owelle finds an igneous rock, which Eric Seagull thinks is valuable. As he carries it, it begins to rain, which erodes the rock. Wind and flooding further break the rock down into sediments. As the sediments sit in mineral fluids, an elephant comes over and steps on it, creating the pressure needed to make it into a (less valuable) sedimentary rock.

Main Ideas

1. Weathering from wind, rain, and floods can break a rock down into sediments.
2. Cementation occurs when mineral fluids act as a glue, helping sediments to stick together.
3. Pressure causes compaction, combining individual sediments into a sedimentary rock.
4. These processes can turn an igneous or metamorphic rock into a sedimentary rock.
5. A rock with sediments that have primarily formed from weathering and erosion is known as a detrital sedimentary rock.

Vocabulary

igneous	sedimentary	detrital
compaction	cementation	weathering

Characters

Owelle, Eric Seagull, and a couple of elephants

Teacher's Note

Point out to students that these processes usually occur over large spans of time. Explaining the role of mineral fluids will also be helpful.
The pressure that causes sedimentary rock formation is typically caused by overlying layers of rock, and not by obnoxious elephants.

Questions for Discussion

Before Reading:

1. What would be the effect of wind and rain on a rock?
2. What makes rocks valuable?
3. Is it possible to change one type of rock into another type of rock? How?

After Reading:

1. How might the processes depicted in this comic occur differently in "real life?"
2. Why was Owelle disappointed by the end of the comic?
3. Why were the mineral fluids important?

Copyright ©2009 by Incentive Publications, Inc., Nashville, TN

Dr. Birdley Teaches Science – Mysteries of the Earth

White Sands

Geology

Objectives

1. To present the gypsum grains of White Sands National Park as examples of chemical sedimentary rock.
2. To explain how chemical sedimentary rock forms.
3. To show examples of natural adaptations.

Synopsis

As Dr. Birdley and Gina walk through White Sands National Park in New Mexico, Birdley discusses the origins of the white sands. Owelle appears, followed by a a group of bleached earless lizards, who are attracted to his smell.

Main Ideas

1. White Sands National Park started as a basin between two mountains.
2. Rain dissolved the gypsum on the mountains and carried it into the basin that would become white sands, forming a lake.
3. The lake evaporated, causing the gypsum minerals to precipitate out of solution, forming the white sands.
4. The gypsum crystals formed by coming out of a solution, becoming chemical sedimentary rock.
5. Wind blew the sands around, forming dunes.
6. An example of a species that has adapted to the white sands is the bleached earless lizard, which has white pigmentation that allows it to blend into its surroundings.

Vocabulary

gypsum	precipitate	dissolve
adaptation	pigmentation	sedimentary
evaporate	pheromone	camouflage

Characters

Dr. Birdley, Gina, Dean Owelle, and a few lizards

Questions for Discussion

Before Reading:

1. What do you already know about sedimentary rock?
2. How could you separate salt water into salt and water?
3. How do you think sand forms?

After Reading:

1. Why is White Sands such a unique location?
2. Why doesn't this type of sand form all over the country?
3. Why are the bleached earless lizards well adapted to the white sands?
4. Do you expect other animals to be white? Why or why not?

OWELLE'S ROCK

NAME: _____

CLASS: _____ DATE: _____

STUDY QUESTIONS

Directions: Answer the following questions to the best of your ability.

1. Was Owelle's rock igneous by the end of the comic? Why or why not?

   _____
   _____

2. Describe the processes that broke Owelle's rock down into sediments.

   _____
   _____

3. A group of sediments are buried under numerous rock layers. How would these layers contribute to the formation of a sedimentary rock?

   _____
   _____

4. Would the natural process of sedimentary rock formation ordinarily be faster or slower than the process depicted in the process? Explain.

   _____
   _____

5. What role do you think mineral fluids had in the formation of a sedimentary rock?

   _____
   _____

 WHITE SANDS

NAME: _____

CLASS: _____ DATE: _____

 STUDY QUESTIONS

Directions: Answer the following questions to the best of your ability.

 1. Describe the significance of the mountain ranges in the formation of the white sands.
_____
_____

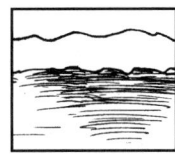 2. Describe how the process of evaporation led to the formation of the white sands.
_____
_____

 3. Describe the role of wind in shaping the White Sands' terrain.
_____
_____

 4. Explain why the bleached earless lizard is well adapted to the white sands.
_____
_____

 5. How is gypsum different from a sedimentary rock that forms from erosion, compaction, and cementation?
_____
_____

A Lesson from a Snail

NAME:_____
CLASS:_____ DATE:_____

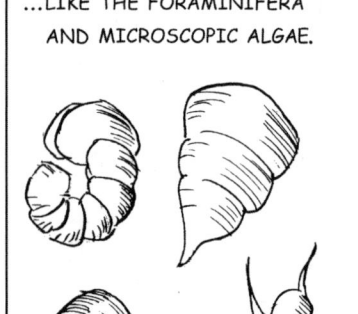

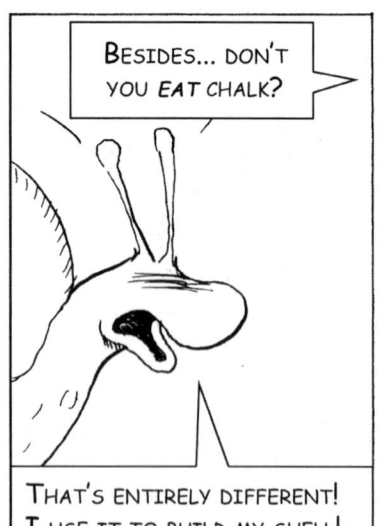

Dr. Birdley Teaches Science – Mysteries of the Earth

A Lesson from a Snail

Sedimentary Rock

Objectives

1. To present an example of biological sedimentary rock.
2. To explain the role of forarminifera and other microbes in the formation of biological sedimentary rock.

Synopsis

A snail criticizes Dr. Birdley for writing on a blackboard with chalk, claiming that he is smearing the remains of fossilized organisms on the board. Norman shows up and defends Birdley, who eventually asks Owelle if the school can obtain any whiteboards.

Main Ideas

1. Chalk is made of a compound known as calcite ($CaCO_3$).
2. Calcite is made from the fossilized shells of marine animals and micro-organisms, such as foraminifera and algae.
3. The foraminifera create skeletons and hard plates which fossilize well.
4. Snails eat chalk because it is needed for the growth of their shells, which are made of calcite.

Vocabulary

calcite	coccoliths	fossils
plates	skeletons	foraminifera
fossilize	biological	sedimentary rock

Characters

Dr. Birdley, Norman, and an obnoxious snail

Questions for Discussion

Before Reading:

1. Where do you think chalk comes from?
2. Do you think micro-organisms are able to form fossils? Why or why not?
3. Why might an animal with a shell eat chalk?

After Reading:

1. Why is the snail angry?
2. Will this comic make you think about chalk differently from here on out? How?

A LESSON FROM A SNAIL

NAME: _____

CLASS: _____ DATE: _____

 STUDY QUESTIONS

Directions: Answer the following questions to the best of your ability.

1. Why is the snail angry at Dr. Birdley for the use of chalk?

   _____

   _____

2. What is the significance of calcite?

   _____

   _____

3. How is chalk similar to the limestone of the Egyptian pyramids?

   _____

   _____

4. Why do you think snails eat chalk?

   _____

   _____

5. Why do you think they consider chalk to be a biological sedimentary rock?

   _____

   _____

SEDIMENTARY ROCKS

NAME: _____

CLASS: _____ DATE: _____

 # BACKGROUND: HOW SEDIMENTARY ROCKS FORM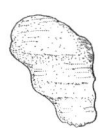

Sedimentary rocks are formed from sediments that have been compressed together. The type of sedimentary rock depends mainly upon the origins of the sediments. The sediments that make up **detrital sedimentary rocks** come from weathering and erosion. For example, an igneous rock exposed to high amounts of wind and rain over a long period of time might broken down into sediments. **Biological sedimentary rocks** are produced in some way by living organisms. Chalk, for example, is formed from the remains of dead micro-fossils that come from microbes such as foraminifera. **Chemical sedimentary rocks** are made of once-dissolved minerals that come out of their solutions. In White Sands National Park, for example, the gypsum sand grains precipitated out of the ancient lake when its mineral-rich waters evaporated.

The Burgess Shale Region in British Columbia is made up of detrital sedimentary rock.

The Cliffs of Dover in England are composed of biological sedimentary rock.

The gypsum deposits of White Sands, New Mexico are a type of chemical sedimentary rock.

Directions: Answer the following questions to the best of your ability.

1. Why might heavy rains and wind result in a detrital sedimentary rock?

_____

_____

_____

2. How is a biological sedimentary rock different from a chemical sedimentary rock?

_____

_____

_____

Identifying Rocks

NAME: _____

CLASS: _____ DATE: _____

IDENTIFY EACH ROCK AS IGNEOUS, SEDIMENTARY, OR METAMORPHIC!

THE PUDDINGSTONE OF HERTFORDSHIRE, ENGLAND IS MADE OF WEATHERED FLINT STONES CEMENTED INTO A SILICA QUARTZ MATRIX.

1. _____

DIORITE FORMS FROM MAGMA THAT COOLS UNDERNEATH MOUNTAIN CHAINS AT CONTINENTAL PLATE BOUNDARIES.

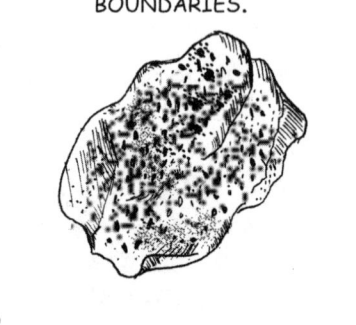

2. _____

PUMICE IS FORMED FROM LAVA FROTH THAT HAS SOLIDIFIED. IT IS HIGHLY POROUS, WHICH MAKES IT THE ONLY ROCK THAT FLOATS IN WATER.

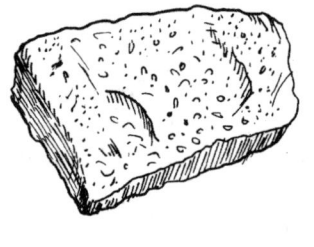

3. _____

SLATE IS FORMED FROM SHALE THAT HAS BEEN TRANSFORMED BY HEAT AND PRESSURE UNDERGROUND.

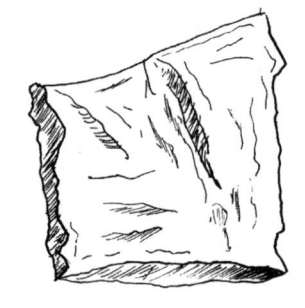

4. _____

GNEISS FORMS AS A RESULT OF EXTREMELY HIGH TEMPERATURE AND PRESSURE IN SUBDUCTION ZONES OR UNDER MOUNTAINS.

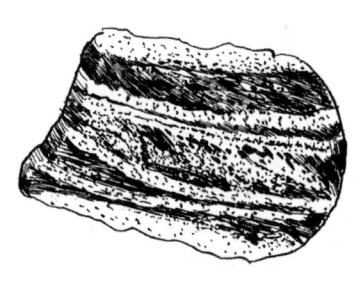

5. _____

BLACK SHALE IS FORMED FROM FINE PARTICLES THAT HAVE SETTLED ON THE OCEAN FLOOR AND THE FOSSILIZED REMAINS OF SEA CREATURES.

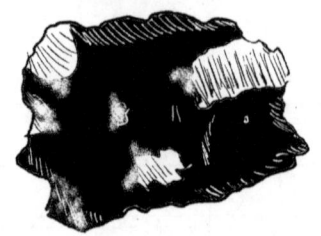

6. _____

MARBLE FORMS FROM LIMESTONE BURIED UNDER THE EARTH'S SURFACE THAT WAS TRANSFORMED BY HEAT AND PRESSURE.

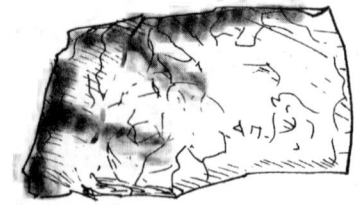

7. _____

SANDSTONE IS FORMED FROM QUARTZ, FELDSPAR, SAND GRAINS, AND OTHER MINERALS THAT HAVE BEEN COMPRESSED TOGETHER INTO HARD ROCK.

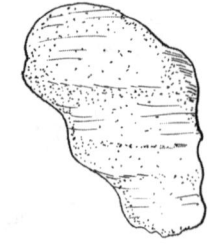

8. _____

SEDIMENTARY ROCKS

NAME: _____

CLASS: _____ DATE: _____

Mini-Comic: Cliffs of Dover

Directions: Read the comic and text. Then, answer the questions to the best of your ability.

Towering above the English channel, the famous Cliffs of Dover were formed over time by the fossilized remains of foraminifera—micro-organisms with hard plates of the mineral calcite, which is essentially calcium carbonate ($CaCO_3$). Foraminifera (pictured in the title panel above) have inhabited the oceans since the distant past. As they died, their hard plates sank to the ocean floor. These plates, known as coccoliths, accumulated over time, and were compressed into a solid rock, which we recognize as chalk.

Their chalky composition makes the cliffs somewhat delicate. Weathering and erosion reduce the cliff face by one centimeter per year, and large fragments have known to fall off without warning. For this reason, visitors are required to stay five feet from the cliff edge at all times. Nevertheless, the Cliffs of Dover remain a massive geological wonder and a significant example of biological sedimentary rock.

1. How were the Cliffs of Dover formed?

   _____

   _____

   _____

2. Why do the cliffs tend to get smaller every year?

   _____

   _____

   _____

 TYPES OF ROCKS

NAME:_____
CLASS:_____ DATE:_____

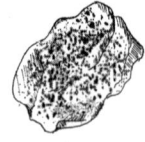 # Graphic Organizer: Rocks

For each major rock category, list its subcategories, explain how it forms, and give some examples. Use the word bank below and feel free to think of additional rocks.

	Sub-Categories	How It Forms	Examples
Sedimentary	1. 2. 3.		
Metamorphic	1. 2.		
Igneous	1. 2.		

WORD BANK

intrusive	erosion	extrusive	foliated	nonfoliated	compaction
heat	pressure	chemical	weathering	lava	magma
detrital	cementation	biological	shale	heating	cooling
sandstone	pumice	slate	gneiss	obsidian	granite
puddingstone	diorite	chalk	marble	limestone	quartzite

Unit 7: Igneous Rocks and Volcanoes

Contents

Source Cartoon: Rocks Above and Below 78

Source Cartoon: Anatomy of a Volcano 79

Cartoon Profiles 80

Study Questions 82

Source Cartoon: Deep Sea Smoker 84

Cartoon Profile 85

Study Questions 86

Background 87

Visual Exercise 88

Dr. Birdley Investigates: Rocks Above and Below

Name: _____
Class: _____ Date: _____

Neil: Dude. I hold in my hand a chunk of obsidian. It is totally *extrusive igneous*.

Let me break it down. It's *extrusive* because it formed on the Earth's surface.

And it's *igneous* because it formed from molten rock.

See, this volcano was spewing out the lava, which cooled off and hardened, forming rocks like this.

The obsidian is all smooth and glossy because the quick cooling kept mineral grains from forming.

Obsidian totally *rocks*.

Dr. Bean: Neil! What happened to Birdley and Gina?

Birdley: Nothing like getting lost in an underground chamber that looks like it's about to cave in. Sheesh.

Gina: Birdley! Shouldn't we find a way to get back to the surface? I think I heard some falling rocks.

Take it easy, Birdley. I just found some intrusive igneous rocks. They're made of hardened magma!

This one is granite. It's *intrusive* because the molten rock cooled below the surface.

And it has lots of mineral grains because it cooled slowly.

Snail: Whoah. That's a piece of my *home*, lady.

Copyright ©2009 by Incentive Publications, Inc., Nashville, TN
Copyright ©2009 by Nevin Katz

Dr. Birdley Teaches Science – Mysteries of the Earth

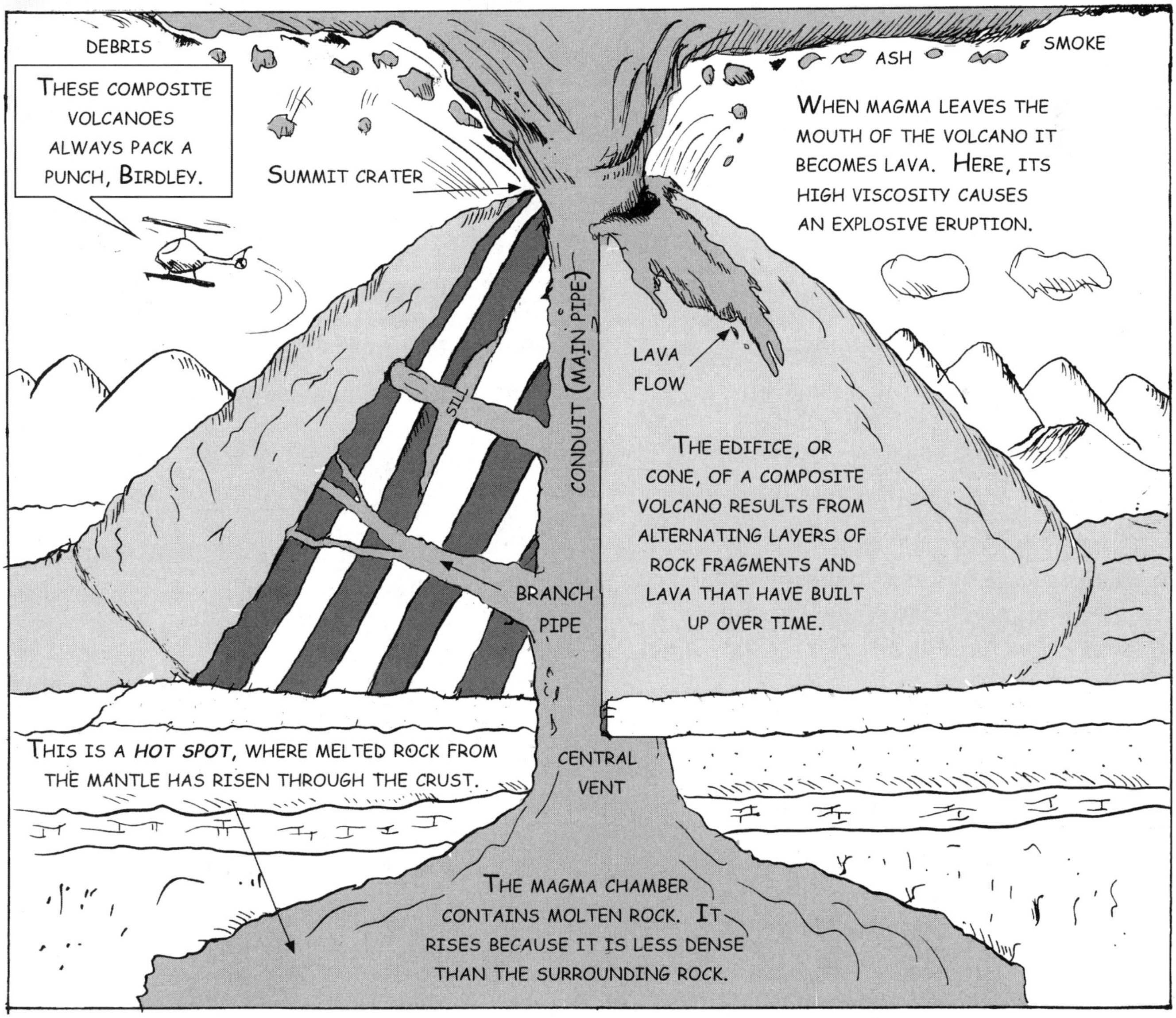

Rocks Above and Below

Igneous Rocks

Objectives

1. To explain the difference between extrusive and intrusive rock.
2. To relate the size of an igneous rock's mineral grains to the rate of cooling.

Synopsis

Neil, standing near an inactive volcano, provides an example of extrusive igneous rock. Meanwhile, Owelle wonders about the location of Birdley and Gina, who have ventured into a hollowed-out rock chamber with Norman. There, they obtain a sample of granite, an intrusive igneous rock, and manage to aggravate an indigenous snail who considers them to be trespassing.

Main Ideas

1. Extrusive igneous rock forms above ground from molten rock that has cooled and hardened.
2. Molten rock that forms above ground is known as lava.
3. Extrusive igneous rock has few mineral grains because it cools quickly, leaving little time for them to form.
4. Intrusive igneous rock forms below ground from molten rock that has cooled and hardened.
5. Molten rock that forms below ground is known as magma.
6. Intrusive igneous rock has many mineral grains because it cools slowly, leaving ample time for minerals to form.

Vocabulary

extrusive   intrusive   lava
magma   igneous   mineral grains

Characters

Neil, Owelle, Norman, Dr. Birdley, Gina, and an obnoxious snail

Questions for Discussion

Before Reading:

1. What do you already know about volcanoes?
2. Where do you think you would find igneous rocks?
3. Why is knowing the location of a rock helpful in determining its identity?

After Reading:

1. Is it possible for intrusive igneous rocks to appear on the surface of the Earth? If so, explain.
2. Would you expect to find extrusive igneous rocks inside the hollowed-out chamber? Why or why not?

Anatomy of a Volcano

Geology, Plate Tectonics

Objectives

1. To illustrate the major parts of a volcano.
2. To explain the causal factors that lead to a volcanic eruption.
3. To illustrate how a volcano might grow over time.

Synopsis

As Birdley is hiking up a volcano, he adamantly declines help from Jaykes, who offers to pick him up in a helicopter. As the volcano erupts, Birdley quickly changes his mind.

Main Ideas

1. A volcanic eruption is a constructive force because it creates new layers of igneous rock that can build up and form structures such as volcanoes.
2. Molten rock from the mantle rises because it is less dense than the surrounding rock.
3. At hot spots, molten rock travels up through the crust.
4. The eruption takes place at the summit crater of the volcano.
5. This particular volcano is a composite cone, because it is made of alternating layers of rock fragments and lava.
6. The high viscosity, or thickness, of the magma allows for great amounts of pressure to build up, resulting in an explosive eruption.

Vocabulary

mantle	magma	summit crater
hot spot	edifice	igneous rock
central vent	lava	magma chamber

Characters

Dr. Birdley, Jaykes

Questions for Discussion

Before Reading:

1. What do you think causes a volcano to erupt?
2. What do you think causes a volcano to form?
3. What areas of the world have volcanoes?

After Reading:

1. How is a volcanic eruption a destructive force?
2. How is a volcanic eruption a constructive force?
3. What type of eruption would you expect from magma with low viscosity?

Copyright ©2009 by Incentive Publications, Inc., Nashville, TN

Dr. Birdley Teaches Science – Mysteries of the Earth

Rocks above and Below

Name: _____

Class: _____ Date: _____

 # Study Questions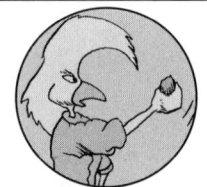

Directions: Answer the following questions to the best of your ability.

1. What is the difference between an extrusive and intrusive igneous rock?
   _____
   _____

2. What do intrusive and extrusive igneous rocks have in common?
   _____
   _____

3. Why do extrusive rocks typically have small crystals and few mineral grains?
   _____
   _____

4. Why would intrusive igneous rocks cool at a slower rate than extrusive igneous rocks?
   _____
   _____

5. Ten kilograms of magma flows through the interior of a volcano. What determines whether the magma becomes an intrusive or extrusive igneous rock?
   _____
   _____

# ANATOMY OF A VOLCANO

NAME: _____

CLASS: _____ DATE: _____

 ## STUDY QUESTIONS

Directions: Answer the following questions to the best of your ability.

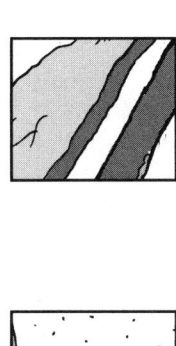 1. Describe how the edifice of a volcano forms.

_____

_____

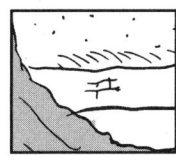 2. Explain the significance of the magma chamber.

_____

_____

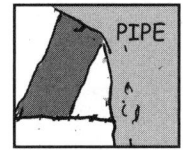 3. Why does the molten rock rise?

_____

_____

 4. Why is the area under the volcano called a "hot spot?"

_____

_____

 5. This volcano is called "composite" because it is made of a combination of materials. Support this statement with a few details.

_____

_____

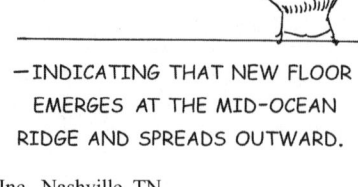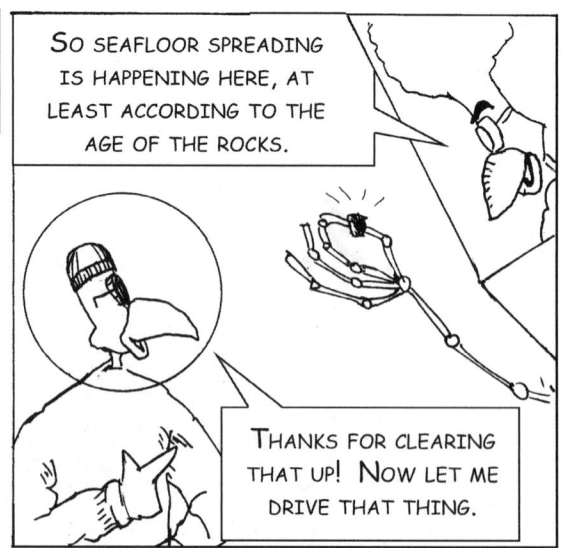

Deep Sea Smoker

Seafloor Spreading

Objectives

1. To illustrate evidence for seafloor spreading.
2. To illustrate the characteristics of an underwater volcano.

Synopsis

Birdley and Jaykes explore a black smoker, which is a volcano that exists on the ocean floor. They meet up with Lark, who has used rock samples from the ocean floor to support the sea floor spreading hypothesis.

Main Ideas

1. Black smokers are volcanoes on the ocean floor.
2. The black smoke they emit is composed of hot water with tiny minerals suspended in it.
3. Black smokers are located on mid-ocean ridges, where magma is rising up through the Earth's crust.
4. As magma rises up through the vent, it cools, forming new sea floor.
5. Microbes live near black smokers and use the chemicals there as a food source.
6. The rocks closer to the vent are younger than the rocks farther away, suggesting that the seafloor is spreading outward from the vent.

Vocabulary

rift valley	extremophile	undersea vent
black smoker	magma	seafloor spreading
volcano	sample	mid-ocean ridge

Characters

Eric Seagull, Dr. Birdley, Jaykes, Professor Lark, microbe

Teacher's Note

The microbe near the volcanic vent is an extremophile that metabolizes hydrogen sulfide.

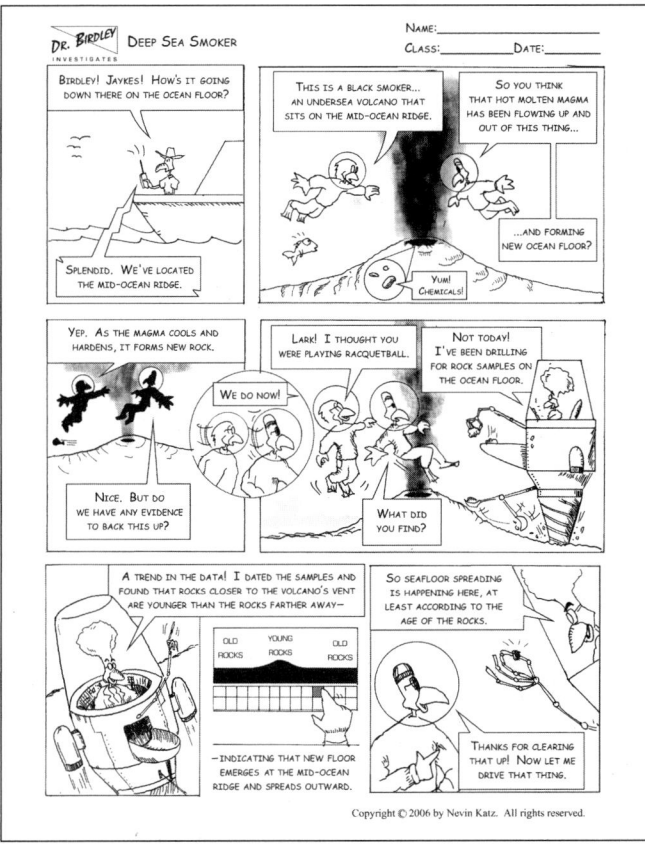

Questions for Discussion

Before Reading:

1. Do you think it is possible for volcanoes to form on the ocean floor?
2. If magma is pushing up onto the ocean floor, where would the youngest igneous rocks be? Close to the vent or far away?
3. Is it possible for living things to exist near the mouth of a hydrothermal vent?

After Reading:

1. How is seafloor spreading similar to the formation of a volcano?
2. Would you ever want to visit a black smoker on the ocean floor? Why or why not?

Copyright ©2009 by Incentive Publications, Inc., Nashville, TN

Dr. Birdley Teaches Science – Mysteries of the Earth

DR. BIRDLEY INVESTIGATES

Deep Sea Smoker

NAME: _____

CLASS: _____ DATE: _____

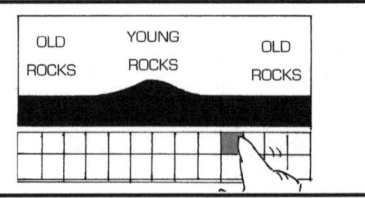

STUDY QUESTIONS

Directions: Answer the following questions to the best of your ability.

1. What is occurring at the mid-ocean ridge?

   _____
   _____

2. How is a new ocean floor formed?

   _____
   _____

3. How is a black smoker similar to a volcano above sea level?

   _____
   _____

4. What abilities would microbes need to have in order to live near a black smoker?

   _____
   _____

5. What evidence did Lark find that supports sea floor spreading?

   _____
   _____

Igneous Rocks and Volcanoes

Name: _____
Class: _____ Date: _____

Background: Volcanoes and Earth's History

Directions: Read the comic and text. Then, answer the questions to the best of your ability.

Despite the ways the Earth has changed over the past billion years, many of the fundamental processes that drive geological events remains the same. For example, consider how volcanoes form. Plate movement and the circulation of magma cause a vent in the Earth's crust to release lava. The flowing of molten rock is constructive in that when it hardens, it forms new rock. The new rock becomes part of the volcano. A similar process happens with undersea vents, where sea floor spreading occurs—except that when the magma cools and hardens, it forms new sea floor. These processes have been occurring throughout Earth's history. In addition to volcanoes and vents, earthquakes and the movement of continents, which are both caused by plate movement, have also been happening for millions of years. As with volcanoes, these processes result from the movement of magma under the Earth's crust.

1. Give examples of geological events that take place today but have also occurred in the distant past.

   _____
   _____
   _____

2. How might scientists know that a given volcanic eruption took place millions of years ago?

   _____
   _____
   _____

Copyright ©2009 by Incentive Publications, Inc., Nashville, TN   Dr. Birdley Teaches Science – Mysteries of the Earth

VISUALIZING VOLCANOES

NAME: _____
CLASS: _____ DATE: _____

LABEL EACH PART OF THE COMPOSITE VOLCANO WITH THE CORRECT NAME! USE EACH TERM ONCE.

PHRASE BANK

MAGMA CHAMBER	BRANCH PIPE
EDIFICE	LAVA FLOW
LAVA AND ROCK LAYERS	DEBRIS
ASH CLOUD	SILL
CRATER	CENTRAL VENT
MAIN PIPE	SEDIMENTARY ROCK

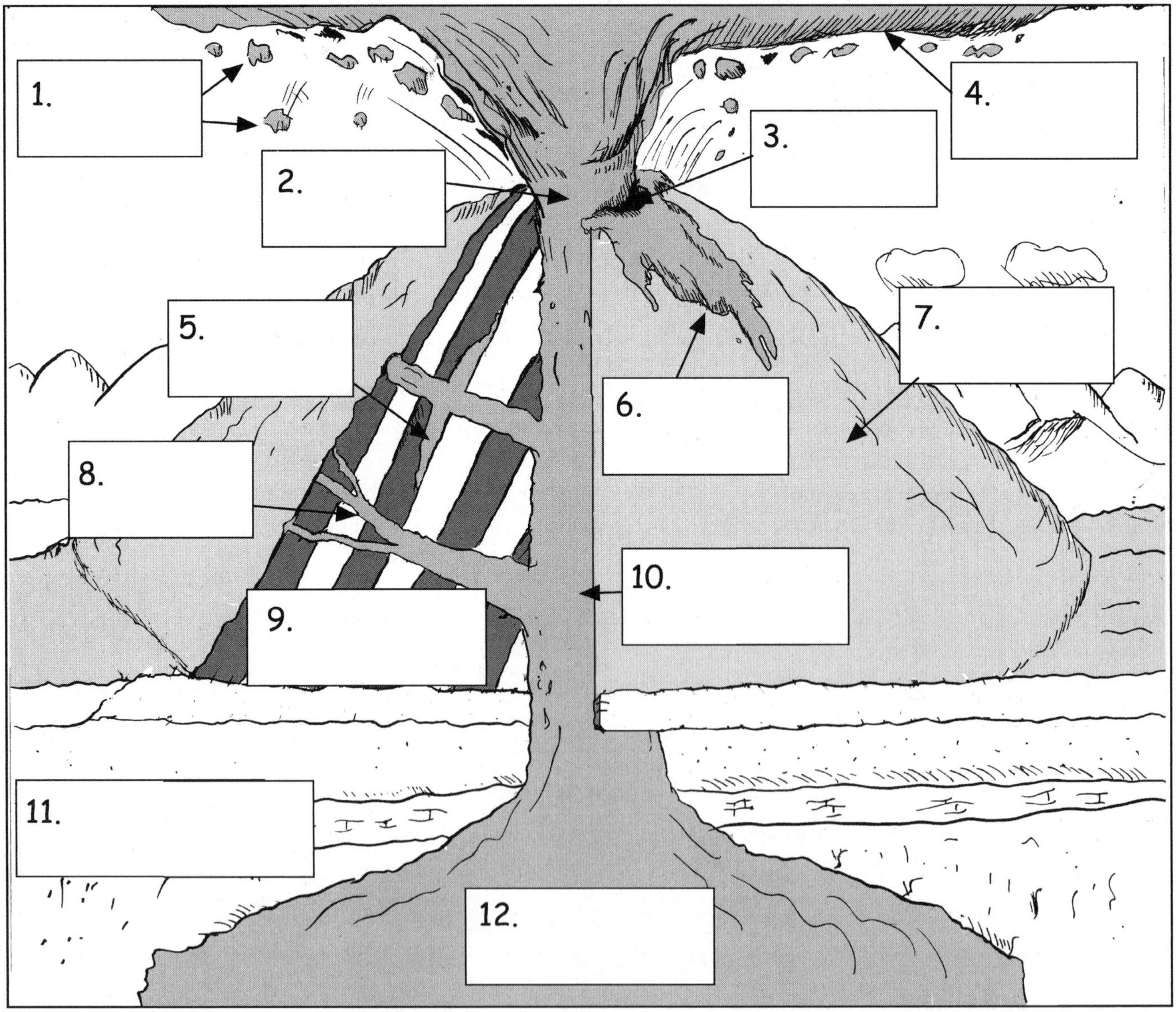

1. _____
2. _____
3. _____
4. _____
5. _____
6. _____
7. _____
8. _____
9. _____
10. _____
11. _____
12. _____

Copyright ©2009 by Incentive Publications, Inc., Nashville, TN

Dr. Birdley Teaches Science – Mysteries of the Earth

Answer Key:

Open-ended questions give science students a chance to make connections between new concepts and their personal experiences, as well as to express their answers in different ways. These questions are one important way to differentiate instruction and address different learning styles. However, open-ended questions do not have a single correct answer, so please read student responses carefully to make sure they are not forming misconceptions. One possible correct answer is always given in this answer key.

page 16: The Fossil Record Study Questions

1. Absolute age refers to the exact numerical age of the fossil. Relative age refers to an imprecise age compared to other fossils.

2. Rocks are important because determining their age enables you to determine the age of nearby fossils.

3. You can only approximate the age of the allosaurus because you can only find the ages of the igneous rock layers above and below the layer containing the fossil.

4. The earthquake could move the fossil layers and cause younger layers to end up below the older layers.

5. The largest age range for the plateosaurus is between 220 and 160 million years ago, because those are the ages of the igneous rock layers above and below it.

page 17: The Fossil Record Study Questions

1. Scientists find a fossil's age by determining how much uranium-235 has decayed in a nearby rock sample.

2. Radiometric dating can be used with sedimentary rocks but not with igneous rocks.

3. The rock has been around for 704 million years, because that is the length of time it takes for half a sample of uranium-235 to decay.

4. Eric used a time frame because he could only determine the ages of the rock layers above and below the fossil.

5. Carbon-14 would be less effective because its shorter half-life of 5,730 years cannot be used with fossils that are over 100,000 years old.

page 18: Searching for Fossils Background

1. Fossils in sedimentary rock may become visible if the rock layers above the fossil are worn away by erosion.

2. Radiometric dating involves determining the age of rocks that are close to fossils. By knowing the half-life of radioactive elements and finding the percentage that has decayed, scientists determine the ages of the rocks.

page 19: Fossil Formation Visual Exercise – see page 93

page 23: Geologic Time Study Questions

1. The timeline indicates that humans have showed up very recently.

2. Microbes have been along for the longest time. Evidence for this includes the oldest fossils of cyanobacteria, which are around 3.8 billion years old.

3. Cyanobacteria and other photosynthetic organisms introduced oxygen into the atmosphere, enabling new type of living things to evolve.

4. Oxygen levels built up first in the ocean and then in the atmosphere. Oxygen-breathers were able to survive.

5. Dinosaurs show up in the later half of Earth's history. The Earth is 4.6 billion years old and the first dinosaurs emerged 235 million years ago.

page 28: Snapshots in Time Study Questions

1. Gina thinks microbes dominated most of the Precambrian because very few fossils of many-celled organisms have been found from that time.

2. Ediacaran time was significant because the first significant fossil records of animals showed up during this time.

3. From the Ediacaran period to the Cambrian period, animal life became a great deal more diverse.

4. Fossils from the Cambrian period reflect an explosion of biodiversity, encompassing all of the known animal phyla.

5. Life primarily existed in the oceans. Perhaps the ocean water contained the nutrients and conditions necessary for early invertebrates.

page 29: The Trilobite Case Study Questions

1. The exoskeleton is preserved because it is made of hard, durable material.

2. In an extinction event, one or more of a species is eliminated.

3. The Permian Extinction greatly affected the biosphere because it resulted in the elimination of 95% of Earth's species, dramatically reducing Earth's biodiversity.

4. A climate change might make the temperature range too high or too low for a given group of species.

5. Answers will vary.

page 30: Background

1. While dinosaurs and smaller mammals emerged during the Mesozoic Era, the Paleozoic Era saw the first fish, first insects, first land plants, and first vertebrates.

2. A plant with with flowers may be better at attracting insects, which would spread its pollen to other flowers.

3. Fossils are more common after the Precambrian because larger animals with hard body parts became more common after this time.

page 31: History of Living Things Visual Exercise - see page 93

page 32: Unit 1 & 2 Quiz

1. b   3. c   5. b
2. d   4. b   6. c

7. Answers will vary, but should explain how the catastrophe would change species' living conditions.

page 38: Origin of Life Study Questions

1. Energy and specific chemicals were required for living things to emerge.

2. Hydothermal vents are rich in organic chemicals and have a continual supply of energy. Microbial communities live there today.

3. An energy surge causes small molecules to self-assemble into larger molecules.

4. Molecules self-assembled into long proteins and strands of genetic material. Then, a functioning set of molecules became enclosed by a membrane.

5. Nobody has replicated the origin of life in a lab.

page 39: Living Fossils Study Questions

1. The microbes are referred to as living fossils because they are similar to the first living things on Earth.

2. Snottites are drippy, gooey projections that hang from the cave ceiling and are largely acidic. They contain millions of microbes that thrive in extreme conditions.

3. It is dangerous to enter the cave because the acidic fluids can be harmful to your skin.

4. This environment is good for the microbes because they are adapted to these conditions. For example, they can consume hydrogen sulfide.

5. Microbes carve out the tunnels of caves because they produce sulfuric acid, which can break down limestone.

page 40: Background

1. Hydrothermal vents are likely candidates for the sites at which life originated because they contain both the necessary organic molecules and the energy.

2. Microbes lived in hydrothermal vents. They thrive because of their tolerance for extreme temperatures and ability to metabolize the nearby chemicals.

page 41: Sulfur Cycle Visual Exercise - see page 93

page 43: Background - Biogeochemical Cycles

1. One reason microbes were important during the Precambrian was that they started cycling fundamental elements through the biosphere.

2. Biogeochemical cycles are processes by which elements travel through living and nonliving parts of the ecosystem, and get incorporated into a variety of compounds. Cycling Elements include nitrogen, carbon, phosphorus, sulfur, and oxygen.

page 48: The Three Domains Study Questions

1. The top diagram in the panel illustrates a classification scheme for living things known as the three domains

2. The many branches included in the bacteria and archaea domains indicate that in some ways, microbes are a great deal more diverse than animals.

3. They are grouped together because they are all more similar genetically than bacteria are to archaea.

4. Microbes are greatly diverse in terms of their genetics, metabolism, and habitats.

5. Because the three domains tree is based on genetic relationships, data on the genetic make-up of living things would be needed to construct the tree.

page 49: Biogeochemical Cycles Study Questionss

1. Microbes consume compounds in the environment and produce new substances. This enables them to recycle elements, incorporating them into new substances.

2. The first microbe consumes oxygen and methane. The cyanobacteria in the ocean produce oxygen, while the swamp microbes produce the methane.

3. Microbes make nitrogen accessible to other living things by capturing it from the air and using it to make ammonium, a nitogenous compound. Ammonium is then converted into other nitrogenous compounds.

4. A biogeochemical cycle is the process by which an element moves through living and nonliving parts of an ecosystem and get incorporated into a range of different compounds along the way.

5. It is important that the world's population of microbes be very diverse because there are so many compounds that key elements end up in - you need a wide variety of microbes to be able to recycle all those substances.

page 50: Carbon Cycle Visual Exercise – see page 93

page 51: Unit 3 & 4 Quiz
1. b
2. a
3. b
4. d
5. d
6. c
7. b
8. a

9. Answers will vary, but may involve microbes that metabolize sulfur-based compounds, produce methane, fix nitrogen, or conduct photosynthesis.

page 53: Vocabulary Build-up, Rocks
1. magma, lava (or the reverse)
2. sediments, compacted
3. transformed, heat, pressure (last two could be reversed)

page 58: Rock Types Study Questions

1. Igneous rocks form from lava or magma that cools and hardens.

2. Sediments may come from erosion and weathering, biological sources, or substances precipitating out of solution.

3. Sediments must be compacted to form sedimentary rock.

4. All metamorphic rocks form as a result of intense heat and pressure.

5. Answers will vary, but may include the following: weathering breaks rock down into sediments. Compaction involves sediments being pressed together to form sedimentary rock. The cooling of magma or lava results in igneous rock. The heating of a rock (combined with extreme pressure) might turn it into a metamorphic rock.

page 59: Transforming Rocks Study Questions

1. Igneous rocks are present around volcanoes because lava, which cools to form igneous rock, flows out of volcanoes.

2. The three stages of rock formation are: weathering and erosion (breaking of a rock into sediments), compaction (the changing of sediments into rock due to pressure), and cementation (gluing together of sediments by mineral fluids.)

3. An igneous or sedimentary rock exposed to intense heat and pressure would become a metamorphic rock.

4. Answers will vary, but may include the following: natural erosion and weathering happen at a slower pace, and are the result of natural forces. Erosion and weathering do not involve mechanical chompers.

5. Yes. An igneous rock could be broken into sediments, which could then be compacted and cemented into sedimentary rock. When exposed to enough heat and pressure under the Earth, the sedimentary rock would then become metamorphic.

p. 60 Background—Types of Rocks

1. Whereas foliated rock has alternating bands of mineral crystals, nonfoliated rock does not have bands.

2. Metamorphic rocks may form at plate boundaries because collisions between tectonic plates can create the heat and presure necessary for their formation.

page 61: Rock Cycle Visual Exercise – see p. 94

page 62: Unit 5 Quiz – Classifying Rocks
1. metamorphic
2. sedimentary
3. igneous
4. b
5. c
6. Shale may be transformed into slate by intense heat and pressure under the Earth.

page 68: Owelle's Rock Study Questions

1. Owelle's rock was no longer igneous by the end of the comic—weathering, erosion, compaction, and cementation changed it into sedimentary rock.

2. Rain, wind, and flooding caused weathering and erosion that broke the rock down into sediments.

3. Pressure from the layers would help to push the sediments together into rock.

4. The natural process would generally be slower because weathering and erosion occur over many years.

5. Mineral fluids serve to glue the sediments together.

page 69: Owelle's Rock Study Questions

1. The mountain ranges had gypsum that was washed into the basin. They also served as boundaries that kept the water in the basin.

2. As the water evaporated, it left gypsum crystals that precipitated out of solution.

3. The wind blew the sand around, forming the dunes.

4. The bleached earless lizard is totally white, which enables it to blend into its environment (camouflage).

5. Whereas some rocks are made from sediments that formed from weathering and erosion, gypsum is made from sediments that precipitated out of a water solution.

page 72: A Lesson from a Snail Study Questions

1. The snail is angry at Dr. Birdley because the chalk is made of the remains of micro-organisms.

2. Calcite is the compound chalk is made of. It is composed of the skeletons of dead micro-organisms

3. Chalk and limestone are both made from the fossilized remains of micro-organisms.

4. The calcite that the chalk is made of is used to build up their shells, which are made of the same material.

5. Chalk is considered a biological sedimentary rock because it is broken down through the biological activity of a snail eating and digesting it.

page 73: Background – Classifying Sedimentary Rocks

1. Heavy wind and rain can wear away a rock, forming sediments. These sediments can then be compacted into detrital sedimentary rock.

2. Whereas the sediments of a biological sed. rock come from the fossilized parts of organisms, the sediments of chemical sedimentary rock have precipitated out of a solution.

page 74: Identifying Rocks Visual Exercise - see page 94

page 75: Mini-Comic: Cliffs of Dover

1. The Cliffs of Dover were formed by the accumulation of fossils from the parts of marine organisms.

2. The Cliffs of Dover get smaller each year due to the weathering and erosion of chalk, which wears easily.

page 76: Types of Rocks Graphic Organizer – see page 94

page 82: Rocks Above and Below Study Questions

1. Whereas extrusive rock forms above the ground from lava, intrusive rock forms underground from magma.

2. Both types form from the cooling of molten rock.

3. Extrusive rocks have small crystals and mineral grains because they cool quickly, leaving little time for mineral grains and crystals to form.

4. Intrusive igneous rocks would cool at a slower rate than extrusive igneous rocks because they form underground, where heat does not escape as quickly.

5. What determines whether magma becomes extrusive or intrusive igneous rock is whether it reaches the mouth of the volcano and spills onto the surface of the Earth.

page 83: Anatomy of a Volcano Study Questions

1. The edifice forms from the build-up of rock fragments and lava.

2. The magma chamber contains the molten rock that will then rise through the central vent and spill onto the Earth's surface during an eruption.

3. The molten rock rises because it is less dense than the surrounding rock.

4. The area under the Earth is called a hot spot because it is a place where melted rock from the mantle has risen up through the Earth's crust.

5. This volcano is made up of a combination of different materials such as lava and rock fragments.

page 86: Deep Sea Smoker Questions

1. At the mid-ocean ridge, lava and smoke is coming out of the undersea volcano.

2. The new ocean floor is formed from the cooling of lava.

3. The black smoker is similar to a volcano above sea level because it is built up by the accumulation of rock that forms from repeated eruptions.

4. Microbes near a black smoker need to tolerate high temperatures and live off of the available compounds.

5. Lark found that the rocks closer to the volcano are younger than the rocks farther away, supporting the idea of seafloor spreading.

page 87: Background – Volcanoes and Earth's History

1. Geological events of the present and past include earthquakes, volcanic eruptions, mountain building, seafloor spreading, and continental drift.

2. Scientists can determine when a volcanic eruption occurred by determiniing the age of the igneous rock that formed from the eruption's lava.

page 88: Visualizing Volcanoes Visual Exercise – see page 94

p. 19 Fossil Formation

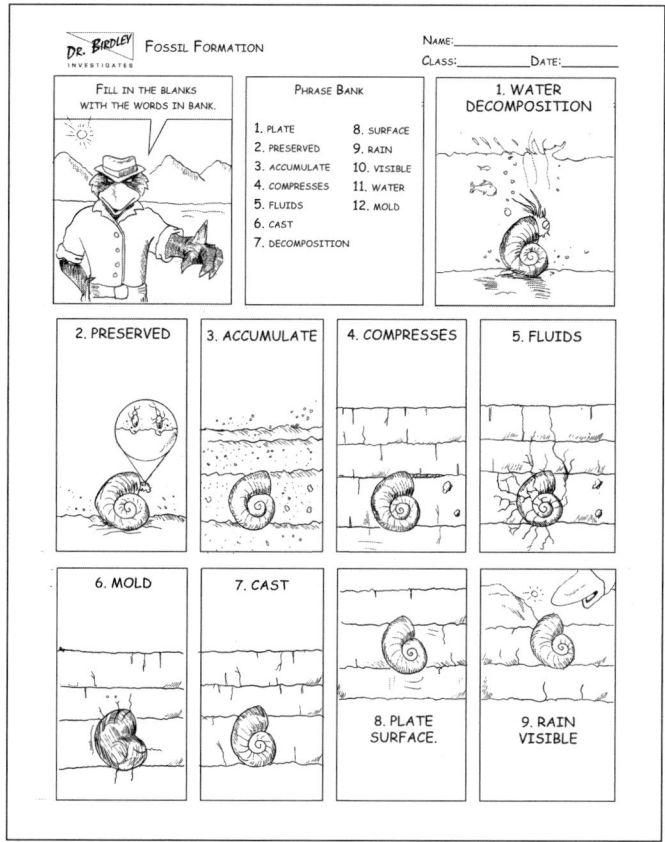

p. 31 History of Living Things

p. 41 Sulfur Cycle

p. 50 Carbon Cycle

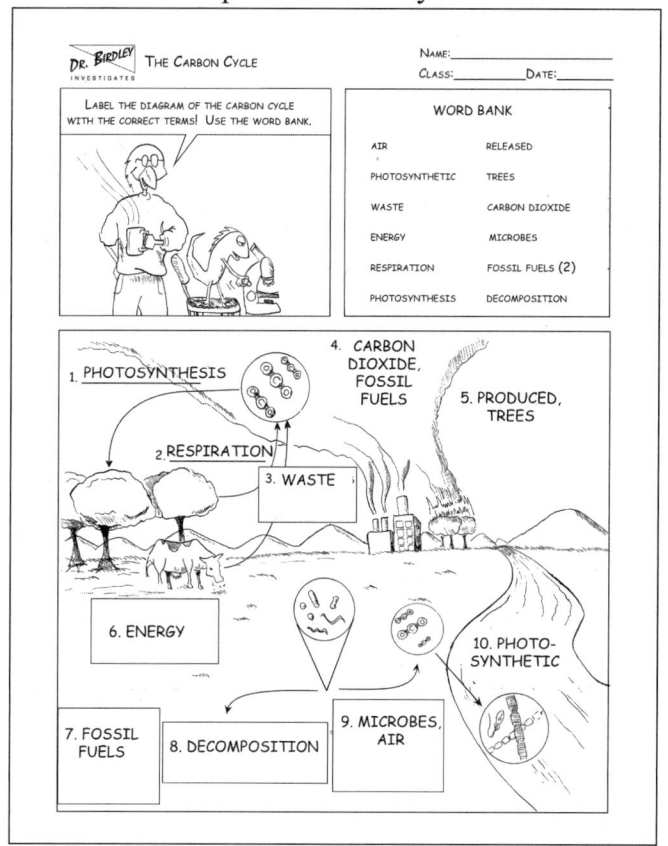

p. 61 Rock Cycle

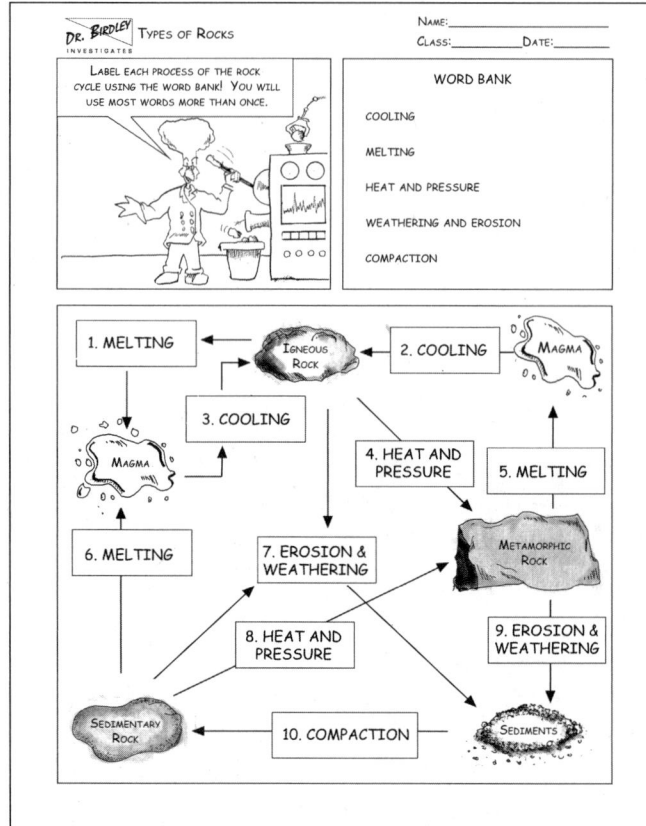

p. 74 Identifying Rocks

p. 76 Graphic Organizer

	Sub-Categories	How it forms	Examples
Sedimentary	1. biological 2. chemical 3. detrital	weathering or other forces make sediments which become compacted and cemented, forming rock	sandstone puddingstone chalk shale limestone
Metamorphic	1. foliated 2. nonfoliated	rock gets transformed due to intense heat and pressure	slate gneiss marble quartzite
Igneous	1. extrusive 2. intrusive	magma or lava cools and hardens	pumice diorite obsidian granite

p. 88 Visualizing Volcanoes

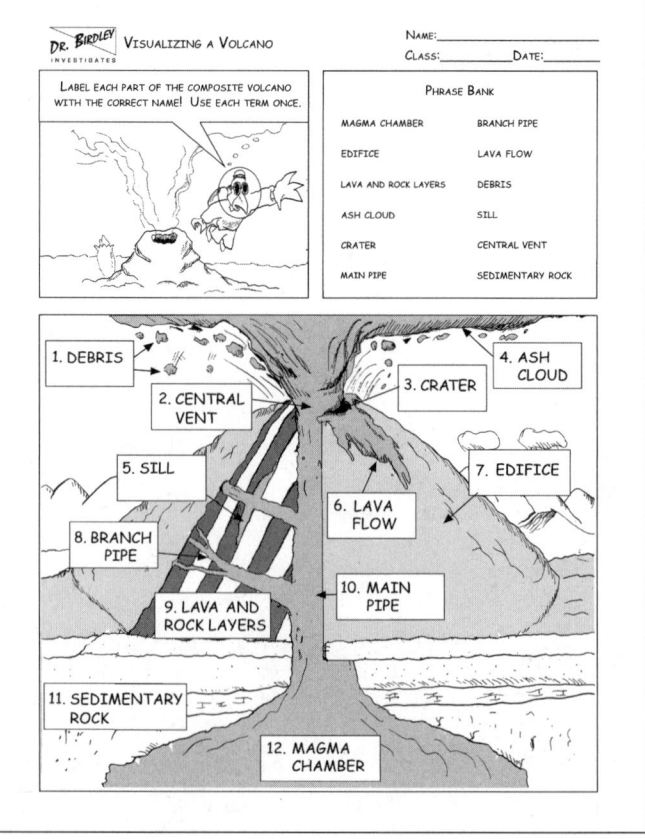

Copyright ©2009 by Incentive Publications, Inc., Nashville, TN

Dr. Birdley Teaches Science – Mysteries of the Earth